Multiple Fire Setters

The Process of Tracking and Identification

Multiple Fire Setters

The Process of Tracking and Identification

Brett Martinez

Dedication

This book is dedicated to my family from whom I have drawn my strength, pride, and hope and for whom I have unbreakable love.

I would also like to dedicate this book to Peter Brennan —— a devoted husband, father, and brother in arms. Pete was a good man in a storm and like all our brothers and sisters who were cut down before their time on September 11th, 2001, shall not be forgotten.

Table of Contents

List of Tables, Maps, and Photographs

ACKNOWLEDGEMENTS

I would like to acknowledge the following people:

Commissioner David Fischler and Chief Vincent Dunn for inspiring me to write this book so that others may profit from my experience in multiple fire setter tracking.

Chief Fire Marshal Warren Horst and Assistant Chief Fire Marshal Peter Digilio for allowing me to experiment and to continue to improve the process of multiple fire setter tracking.

Fire Marshal John Coen for his command of the English language.

Many of the maps were produced using "Undertow Software, Inc.'s Precision Mapping software, © Copyright 2002."

INTRODUCTION

Since the 1990s the field of crime analysis and crime mapping has steadily grown at the local, regional, and state level. These units have focused on the most noticeable repeat offenders, which include serial rapists, muggers, and burglars. Most have been successful in hunting these violent offenders. Unfortunately, most have not focused on or even attempted to hunt serial arsonists. These violent offenders, along with serial killers, have been the focus of federal assets employing a different technique.

The Federal Bureau of Investigation (FBI) has estimated that there is an average of 40 to 50 serial murderers operating in North America at any given time. Although it is an assumption, it would be safe to estimate that there are an average of 50 or more multiple fire setters (MFS) operating in North America within the same given time. This estimate is derived by using research techniques similar to those used by the FBI. One of the research techniques the FBI uses is to check newspaper reports on violent crimes. By using newspaper reports I was able to identify numerous incidents throughout the US and Canada where patterns became evident. I was able to access these sources by going online with my computer and finding a search engine for news media and then typing in "arson". The cases could be found in the smallest communities as well as the largest. In one article the chief of fire management programs at the United States Fire Administration stated that "pyromaniacs" account for approximately 14% of all intentionally set fires. My research showed the numbers were higher than I would have imagined.

With the proliferation of national news reporting services and the increasing need for bigger stories, the serial criminal has become one of the most popular lawbreakers for attracting media attention. Due to this increased media coverage and the need to cover every angle of the story, the interest in people who hunt the offenders has become as great as the interest in the criminals themselves. The technique used by those who hunt the criminal is called profiling.

Profiling uses a combination of the forensic scene investigation, along with a psychological study of the criminal mind and analysis of criminal activity in an effort to capture violent offenders. The proliferation of information on profiling can be found in many places, including clinical studies, novels, television shows, and even movies about profiling. In the field of fire investigation, the profiling of MFS has been led by the Federal Bureau of Alcohol, Tobacco, and Firearms (ATF), and the FBI at the National Center for the Analysis of Violent Crime (NCAVC) in the Virginias. Their success in this area has been well documented and we can expect continued growth in this field by these and other law enforcement organizations.

However, the following was not written to discuss profiling methods or the forensic psychology used to investigate and interview MFSs (multiple fire setters). Instead, the purpose of this book is to provide an objective technique for the identification, tracking, and eventual apprehension of MFS by using basic information available to those willing to access it. This information is usually accurate, timely, accessible, and most important, free to investigators. The knowledge on which tracking is based has been developed through collecting, plotting, and analyzing information from numerous sources. Through working and training with investigators from jurisdictions all over North America, the process has expanded to include informational sources that have been overlooked in the past. The accumulated data from these sources has allowed the investigator in the field to be well informed and ready to anticipate the offender's next move.

This book is not designed to replace the profiling analysis process nor crime analysts, although forensic psychologists and profilers may find it useful and interesting. Instead, its primary purpose is to aid fire investigators in identifying MFS. It should be regarded as a tool to advance the investigation process. The first step in the profiling process should be to identify the activity of the serial offender/MFS. Documented evidence of their consistent activity or offenses can advance the tracking, crime analysis, and profile development that aids in their apprehension. This, perhaps, may even lead to catching them in the act. For offenders who do not seem worthy of the profiling effort, this procedure will also help to apprehend those who will probably never graduate to more than "firebugs". Firebugs are individuals who frequently appear in many communities. Typically they are known by names such as "Lenny the Light", "Bicycle Bob", and "Tony the Torch". They often achieve early release from prison, rarely get prosecuted for anything more than petty offenses, and always return to their old habits in the same community upon release from incarceration.

This text can additionally be used as a case study in investigative techniques for fire investigators or as an aid to crime analysts and profilers. However, it is designed in such a way so as to serve as a manual for a broad classification of professions associated with fire investigation, each with varying concerns. Fire investigators, fire prevention officers, police officers on patrol, fire department line officers, and firefighters will find its varied and multiple applications useful.

This book will be equally useful to fire prevention officers when dealing with juvenile fire setters as well as others. After conducting a canine demonstration for prospective criminal justice students at the New York Institute of Technology, I had the opportunity to discuss my theories with a professor in the area of criminal behavior. I then realized that this book could also be extremely useful to the newest type of crime fighters–crime analysts or crime mappers. These crime analysts and their civilian counterparts, the GIS (Geographic Information Systems) analysts are on the cutting edge of crime fighting, and can use the information in this book as a guide for the mapping of fire activity. As mentioned earlier, most crime analysts have had limited experience in hunting MFS for varying reasons. Whether it is due to limited fire investigation experience, unfamiliarity with the MFS methods, or some other reason, the fact is that MFS can be analyzed and identified. It may take a little more effort at the start of the analysis but it will pay off in the end. The efforts of crime analysts will provide tremendous amounts of information to all concerned with the MFS. For example, the basic foot soldier, the patrol officer, who is assigned to a patrol area will realize that simply observing and noting the license plate number of a vehicle in the vicinity can be extremely significant. The fire officer and firefighter will find it useful in recognizing the indicators and clues that are the signature of an MFS, thereby resulting in the development of appropriate strategies and tactics to prevent further occurrences. It will also help to differentiate between patterns of MFS and trends in standard fire activity.

This book will not make any one of these individual professionals an expert in any of the other professions mentioned, but it will serve to refine and enhance their expertise in the areas that they specialize in. Hopefully this book will help members of all the professions mentioned gain a better appreciation of what goes into an MFS investigation. This appreciation will help all these involved professionals not only understand the other parties' roles, but also see more clearly appropriate procedures for each, before, during, and after the investigation. It will also define what *not* to do.

For those unfamiliar with fire investigation or major cases, this will be vital knowledge to attain. Unlike most investigations, a fire case will involve multiple agencies. The two most obvious groups would be the fire department who extinguishes the fire and the police who investigate the fire. However, at a fire scene these two groups do not always have the same goals in mind. For instance, there is frequently a lack of and/or miscommunication pertaining to the goals of fire services as opposed to those of law enforcement investigators. Too often, investigators have an incorrect perception of what the focus of other individuals or groups is during the investigation. Too many times the goals of one conflict with the goals of another simply because no one has bothered to explain, understand, or properly communicate their intentions to the other organizations involved. Due to this situation, a valuable portion of time is spent mediating in order to facilitate cooperation between the two conflicting agencies. A case in point is the 1995 TWA Flight 800 crash off of East Moriches in Suffolk County, Long Island. Before the crash, it was wrongly believed that the disciplines would cooperate and work together at an aircraft crash scene. However, less then six months before the crash of Flight 800, a single engine aircraft went down in a heavily wooded state preserve. The passengers were presumed dead by the police (making it a homicide investigation) and presumed missing by the fire and rescue services (making it a search and rescue mission). The commanding officer (CO) of the homicide squad actually attempted to prevent fire and rescue personnel from responding to the scene, by standing in the path of fire apparatus entering the wooded area. If the CO had not moved at the last minute, he would have been run over, numerous times, by responding apparatus.

With this in mind, this book has been organized so that every chapter can be referred to when encountering a specific issue related to tracking. For example, if you need to convince your superior why tracking should be done and why they should cooperate with other agencies, simply read chapters 1 and 3. If you need to know what is required in equipment specifications, refer to chapter 11. This manual will provide all professions mentioned with the ability to develop the tools needed to help stop and prevent multiple related fire incidents.

Many of the concepts used in this process are not new. Throughout the United States, investigators have been using some of these ideas in their routine investigations for years. Public safety agencies all over the US have begun using computer software for crime analysis with varying success. In some cases, these analyses have helped to develop anti-crime operations over a broad scale and during some major cases, investigators have used most, but not all, of the concepts outlined in this book. My research has indicated that never before have all of these concepts been consolidated and formulated into a process that can be used by any of the professions mentioned to track MFS activity. This would not have been possible without the aid of informational sources on fire incidents, which are vital to initiating and updating the tracking process.

In the nine years just prior to the writing of this book, I have helped to track multiple cases concurrently. During that time, we have identified at least 17 MFS cases involving hundreds of fire incidents and dozens of malicious false alarm calls. Ten of these cases have led to arrests, three are closed with the fire activity ceasing and the rest remained open or active investigations. Many other incidents that we began to investigate have stopped entirely when word spread through the community that fire officials were perusing patterned fire activity that was occurring in the area.

I have focused on the process of analyzing and refining the information gathered from these multiple sources and over the past nine years it has been continuously refined and updated. As the process becomes more efficient, less time is being spent collecting information and thus allows for more time for tracking the fire activity of MFS. The end product of this process is a report, simply referred to as the Fire Incident Tracking System or FITS Report, which is basically a tip sheet. This report is concise and designed to be used by units in the field, with the information available to all investigators and all concerned parties to varying extents.

Limited information related to each profession's responsibility and investment in the case should be shared. However, keep in mind that not all the relevant information needs to be divulged to each person. By allowing a somewhat free flow of information in this manner, you can create an atmosphere in which those who receive the information may provide significant other information that is vital to tracking the MFS.

Currently, the process has been refined to a level that only requires a few hours of effort a week for search and analysis since the information and reports are all computerized and can be quickly modified. The significance of this technique is important enough to share with fellow investigators. There will be little or no distinction made between the types of fire setters (juvenile, serial, pyromaniacs, etc.). The reason for this will be explained in detail later, but suffice to say for now that the tracking process is not designed to show distinctions between the motives or types of individuals who set fires. A trained profiler or forensic psychologist would recognize certain distinctions and/or behavioral characteristics about the offender from this information. Properly trained professionals will find this information, along with evidence, crime scene reports, and photographs, useful in developing the unknown subject's (unsub) profile.

Future chapters will discuss how this information can be useful to profilers, crime analyzers, crime mappers, and investigators, but for now remember that this process is designed to aid those professionals who deal with fires. It is designed to track fire activity as well as help catch MFSs. There will be a brief discussion on motives, but nothing that will make you a specialist in this field. For more information on that subject and other fire investigative concerns, the bibliography and glossary of this book will contain sources that can be used for further research. Those research sources, not only helped me develop this book, but also enhanced my understanding of the fire setters we are tracking.

This book will focus on tracking all MFSs regardless of their motivations, and as such, will not be a behavioral science study. MFSs have a pattern to their activity that they are not even aware of. Recognizing this pattern provides the investigator a tremendous advantage. When Sun Tzu stated in *The Art of War*, "know your enemy", he not only meant psychologically, but also his or her routine, schedule, and order of battle. This process will help you understand the strategies of MFSs and you will use it to identify an individual fire setter's tactics. Because MFSs set fires based upon time, availability, and opportunity of target, they unknowingly give the investigators or trackers the information needed to establish surveillance, to conduct interviews, and to provide probable cause to search. With time, your experience in tracking will enable you to pick out which incidents are related to the offender being tracked and which are not related, simply by reviewing the fire activity logbooks. You will be able to develop reports and alert fellow professionals to pending activity. You will be able to develop maps and data that can be used by interviewers and prosecutors. You will have the tools to improve this process and better understand the unsubs (unidentified subjects) you are tracking.

Chapter 1

WHY BOTHER TRACKING SOMEONE?

Before we get ahead of ourselves, let us discuss the reasons why we should track fire activity and the concept of tracking. Some will ask, "Why bother with tracking these minor fire incidents at all?" This is especially true because most multiple fire setters (MFSs) set nuisance fires (*i.e.*, brush, dumpster, rubbish, and abandoned structures). Unfortunately, the laws established to punish individuals for setting fire to unoccupied structures or any of the others examples mentioned have historically been weak. When there are arrests, they are written up as misdemeanor offenses, which are pled down to violations with no jail time.

Many times the laws that would address these crimes are unknown to local law enforcement officials and prosecutors. In some cases there are no laws addressing the wanton act of igniting materials or property. This problem exists in many jurisdictions, where few laws are created to address the problem of the burning of grasslands, rubbish, or old tires. Unfortunately, these three types of fires have been the handiwork of MFSs, causing severe economic losses to communities, as well as serious injuries and even fatalities, to those who extinguish them. For example, in Philadelphia in March of 1996, a tire fire set by serial arsonists caused the shutdown of Interstate 95 for two days, thus crippling the main transportation line for interstate commerce in the Northeast. The resulting economic loss was enormous. In another case a firefighter was severely burned by a fire set by a MFS (known as the "Northeast Side Arsonist") in the late 1990s. The suspect is believed to be responsible for setting wildfires consistently, during the month of August in Kern County, California. In 1983, rubbish fires set by juvenile fire setters in a New Jersey city caused a two-day conflagration that resulted in the evacuation of neighborhoods and the destruction of businesses and homes. The suspects were charged as juvenile offenders. News reports considered them misguided individuals, who have since been released.

When suspects are arrested, they are usually written up according to laws that do not accurately reflect the crime. Because criminal mischief is not a severe enough charge prosecutors seek indictments that will result in more harsh punishment in order to satisfy the public outcry. This only leads to appeals by the defense as excessive charges or misuse of justice.

In other cases, fire starters are charged with criminal mischief that does not adequately reflect the seriousness of the crime. This occurs because law enforcement officials lack sufficient knowledge regarding laws that would more appropriately suit the offense. Local, state, and federal law enforcement officials must work together to overcome these problems by establishing task forces and law enforcement councils. The formations of these groups will allow the various agencies to sit down as equal members, share information and determine goals. These councils can then work to design legislation to combat the problems and then educate law enforcement officials and prosecutors regarding the new laws. With more appropriate laws in place and the personnel properly trained, agencies can continue to work together to develop strike forces to apprehend the criminals. For example, in Suffolk County, NY, an Arson Task Force was established in the early 1980s to educate, share information, develop funding, and combat the crime of arson. In the 1990s the Pine Barrens Law Enforcement Council was formed in Suffolk County. This council deals with all law enforcement issues related to protected lands. Currently, in a joint effort, both of these groups are lobbying for the adoption of a state law that would identify any subject who has intentionally set fires as an arsonist. This would expand current laws to include grassland, abandoned auto, rubbish, and rubbish container fires. The task force concept has been very successful over the last two decades in catching, not only arsonists, but also murderers and bombers. Successful detection and prosecution should be the long-term goals of any agency. For the short term, the tracking of fire activity can be used to help convince lawmakers of the magnitude of the fire problem. For now let us continue to focus on the short-term goal and answer the question posed earlier: Why bother tracking fires?

The answer to this question is twofold. First, most MFSs will not be satisfied with just setting one rubbish or small brush fire, but will progress or graduate to larger, more destructive structure fires, always looking to push the limit without getting caught. It is always better to catch them early even if for a small fire because they are then "in the system." In some states, fire investigators I have worked with and spoken to, have told me that the names of subjects in case files can be kept in a database file such as the Arson Information Management System (AIMS). Names and personal information are then accessible and can be disseminated. However, it is strongly recommended that legal counsel be consulted before pursuing this type of information file. This is based on the fact that a previously interviewed subject or suspect that is continually look upon as the potential offender in unrelated cases, may consider this harassment and it could lead to complaints against the investigative agency. Additionally, if a person remains only a suspect and not convicted, there are legal limits as to how extensively such information can be used. Any files kept on that subject can be considered biased and can contribute to the formation of a smear campaign. In some states there are even laws prohibiting government agencies from keeping files on citizens based on hearsay.

Secondly, when you identify fire patterns early on, the opportunity to capture the offender is greater because in most cases the MFS starts small and is just learning the trade. Such a subject will make the majority of his or her mistakes while in the novice or beginner stages, making it easier to identify them. For instance, simply cross-referencing burn reports with fire incidents may bring success. We have had success in checking burn reports after suspicious fires have been started with flammable liquids, since the possibility of the suspects receiving burns are high. This is especially true for suspects who underestimate the low order explosive nature of many flammable liquids.

The MFSs also start their careers by setting fires in locations where they feel most comfortable and safe. This area will usually have some significance to the suspect. Certain types of MFSs will continue to return to this location. These areas are known as the "cluster centers", "hot spots", "crime clusters", "cluster zones", and "ground zero" and may provide clues as to the identity of the arsonist. By identifying the pattern early in the fire setter's career, ground zero can be located and used by investigators, profilers, mental health experts, and prosecutors. As mentioned earlier, this is the time the fire setter is most vulnerable to detection and apprehension. Useful clues may be in plain site of the original cluster center incident. The fire setter is trying to build up enough confidence to escalate the violence, or in some cases, is reaching out for help.

When the cluster centers are identified and searched, additional patrols will become the normal standard. Some MFSs will begin to notice increased police presence and in some cases will stop before being confronted. Others may stop upon being confronted. For example, with some juvenile fire setters and new firefighters a simple confrontation with law enforcement and fire personnel is enough to scare them off. By confrontation I mean inquiring, "Do you know who started this fire?" or "Did you start this fire?"

Other MFSs will continue to set the same type of fire over and over. This may not seem to be dangerous until the first firefighter or civilian is hurt or killed, at which time it will be regarded as a major threat. In these cases the fire setter may assume that no one noticed these fires. When someone starts to pay attention, they are scared off and the fires stop. From a fire official or fire prevention officer's point of view, this can be just as effective as an arrest, for the fires will stop. When a community is being terrorized, the residents, the leaders of the community, and the media all tend to focus on one thing—making the fires stop. That does not always mean an arrest. The victims will want justice, but the community wants or demands that the fires stop. Consequently, the ultimate goal of this process is to have the fire activity cease.

If these reasons have not satisfied the need for continuing with the process, consider the following: What do Ted Bundy, David Berkowitz, and Charles Manson all have in common? In addition to being serial killers, they were all MFSs. According to Special Agent Anthony Olen Rider's (Behavioral Science Unit FBI, 1980) study of Fire Setters, David Berkowitz set over 2000 fires and made 137 malicious false alarm reports in New York City from 1974 through 1977. Bekowitz even called himself "the Phantom of the Bronx".

Obviously, these subjects are individuals suffering from true mental illness. They have a true propensity for violence, falling into the homicidal triad category where fire setting is just one of three traits. The other two are cruelty to animals and bed wetting into the adolescent stage. The incidents I am speaking of are those that receive little or no criminal punishment. They may be referred to as juvenile curiosity seekers, mischievous children, or individuals searching for help. They are still in the initial stages of experimentation looking for attention and in the case of firefighters, looking for action. If the answer to the question mentioned earlier ("Did you start the fire?") is yes, then further action must be taken. Referral to a mental health expert is probably the very least that should be done. If they admit to the act, especially juveniles, then there is a good chance they are reaching out for some kind of help. Irene Pinsonneault, an expert on juvenile fire setters, said it best: "The one thing that all incarcerated arsonist and institutionalized pyromaniacs have in common is, they traced their fire-setting history back to when they were 3 and 4 years old and they were just playing with matches." Since it is not uncommon for serial killers to have been multiple fire setters, knowledge of their previous patterns of setting fires might have aided officials in identifying and apprehending them sooner and even saved some lives. This would appear to be another important reason to track MFSs.

The thought of tracking down some suspect may conjure up images of John Wayne ("The Duke") in a John Ford western, acting as a deputy US marshal, hunting down some outlaw through the plains of Texas. Even though the days of literally tracking a suspect through the open plains has disappeared, tracking suspects through rural, suburban, and urban areas continues. This idea of tracking fire activity is nothing new. The first documented studies date back to the late 1970s when law enforcement agencies attempted to pattern search for fire activity. Additionally, Dr. David Icove wrote "Principles of Incendiary Crime Analysis: The Arson Pattern Recognition System (APRS) Approach to Arson Information Management," Kentucky Department of Public Safety, 1983. This was the landmark work for fire tracking that continues to be studied today. Components of this work can be seen in this process and in Orion software referred to as Geographic Profiling or Geoforensic Profiling, developed by Dr. Kim Rossmo of the Vancouver Police and Environmental Criminology Research Inc. At that time the process to get the information needed to conduct crime analysis (using data entry and retrieval) was, at best, cumbersome and required the help of expert computer programmers. Fortunately, even though these early attempts at tracking fire setters were somewhat tedious, the concept did not die, but instead continued to grow. This crime analysis process has had additional success in many cities throughout the United States. For example, the city of New Haven, CT, pioneered this process during the 1980s. A similar process is now used successfully in many other cities for tracking arson for profit (to a lesser extent) tracking properties with fraud potential, and preventing future fires.

Although early on it was not easily accomplished, tracking has continued to prove its value to investigators. Tracking has been used in many high profile cases such as the Paul Keller case in Washington State and the John Orr case in California. Both Orr and Keller set fires along the Pacific US in the early and mid 1990s. The investigation of them and their arrests are well documented. The arrest and conviction of both individuals is a testimony to the cooperation and joint efforts between federal, state, and local law enforcement agencies to catch MFSs.

The difference between those cases in the mid-1990s and now is primarily the continued advancements in technology. The naval scholar Mahan said, "technology dictates tactics". This rings true not only in naval warfare, but also fire fighting and fire investigation. Technology now allows you to track MFSs in and around your jurisdiction. The key to successful tracking is to have user-friendly technology. In this book we avoid using terms that do not enhance this idea of user-friendliness for they have no application in the field. Instead, we use "real" world terms (such as linking incident activity) as opposed to "virtual" world terms.

In the past, only large losses or high profile cases were deemed worthy enough to be tracked. This was the situation when I began tracking. We began by backtracking and looking at prior incidents when an arrest was imminent. After the arrest we began looking at other incidents that our suspect may have been responsible for. We then realized that our suspect did not start his fire setting career with structures. What had actually occurred was that the suspect (at that time the unknown subject) started out small, with rubbish and small, dry vegetation or brush fires. At this stage, our unknown subject (unsub) experimented with different ignition sources. Following the brush fires the unsub began setting fence fires and igniting small structures such as sheds. These allowed him the opportunity to develop the skills of target assessment, ignition tactics, and evasion techniques. Soon after, the unsub began testing all of these skills on abandoned structures. When this was not satisfying enough, the unsub moved to new construction and occupied buildings, escalating the violence with every fire. Unfortunately, by the time we noticed that we had a serial arsonist, the unsub was setting multiple fires in large structures simultaneously. Our suspect had been setting fires for almost a year before he was identified. Toward the end we were playing catch up, while allocating tremendous resources, such as undercover units, helicopters, K9 units, detectives, and fire marshals working all hours of the day on the case. Before the suspect was finally apprehended, he had set four more fires, killing two people and successfully terrorizing an entire community.

After the knowledge we gained from that case in 1991 and two other serial arson cases in 1992, we began working with "profilers" in Virginia. This was very helpful and was the catalyst for the development and training of fire personnel to combat firefighters who had become arsonists. At first we believed it was only volunteer firefighters and juveniles who had fallen into a pattern of setting fires. Quickly we realized that this was inaccurate and that even full-time firefighters sometimes fall into the same pattern. Although we continue to investigate firefighters who intentionally set fires, the number of cases has dropped dramatically. Currently, firefighter arson cases only account for a small percentage of multiple fire setters, but almost all lead to arrests.

In 1991 when I first dabbled in the process of tracking fires, the gathering of information was all legwork. By 1992, the gathering of information was still done by hand, but I had begun to locate and use other resources that were available by phone. Once I had established these information resources, the next step in developing this process was to analyze the information gathered and determine what information was most useful and what could be discarded or set aside. This was a trial and error process. The information gleaned from combing through fire department logs and run sheets, and the cross-referencing of dates and times, required that they all be plotted manually. The most sophisticated technological items during this time were the photocopy and fax machines. The results were satisfying, but because it took so long to gather and analyze, the data only helped to confirm our theory about fire setting patterns of a suspect already under arrest. The procedures for gathering and organizing pertinent information needed to be developed and refined.

The transfer of information to databases evolved next. Converting the information to data, analyzing the data, plotting the results, then re-analyzing and disseminating the results was still a slow or even nonexistent process. Over the next few years, new resources were developed and cultivated. Now, increasingly sophisticated technology has become more affordable to governmental agencies and more user-friendly to the investigators involved. The use of technology to track most, if not all MFSs, on a daily basis is now a reality.

Chapter 2

HOW DO WE DO IT?

Once again it needs to be stressed that reading this book will not make a profiler out of you. Nor will the mastering of this process alone train you as a fire investigator. These two specialties will require much more training and many more years of field experience than could be offered in a few paragraphs (for that kind of training, checking with the local fire and/or police academy, the National Fire Academy, or the Federal Law Enforcement Training Center is recommended). This book will not even make you an expert in the use of profiles. No fire is the same; therefore, no one can be an expert on all fires. What this book will do is give the investigator (or tracker) the knowledge and basic skills to track and look for patterns in fire activity. Every field requires a specialized vocabulary essential for clarification. Likewise, before we can track, we need to know a little more about some of the terms and definitions associated with MFSs, as well as the types of people that will be tracked and their motives. Many of these terms are those explained by Lewis and Yarnell in their book, "Pathological Fire-Setting (Pyromania), Nervous and Mental Diseases Monograph" published by College Foundation, New York, 1951, AO Ryder.

Please note that in the text I refrain from using the term *arson* too often. The definition of this word will vary among state and federal jurisdictions. Due to this variation, I will continue to avoid the word whenever possible. Instead, I will use the terms *set* or *ignited fires* and will define them to mean the following: any intentional fire act with the purpose of doing bodily harm to others or with the intended purpose of destroying and/or damaging objects and/or property. The words *incendiary* and *suspicious* will be used when discussing MFSs. Incendiary will refer to any device used to start and accelerate a fire, while suspicious will be used to explain any fire that is not natural or accidental in origin.

Lay terms and names associated with MFSs have included *pyromaniac, pathological fire setter, compulsive fire setter, firebug,* and, more recently, *serial arsonist* and *juvenile fire setter.* For the purposes of this book, all of these different types and terms will be referred to as *MFSs* or *MFS* because they all share some common traits.

Although I am willing to venture a guess regarding the psychological similarities, we will concentrate on the behavioral patterns of MFSs during and after the fire incidents. A list of these similarities follows:

- All MFSs learn their craft and ignite fires according to a schedule, usually theirs.
- They all "like" to set fires, not one or two, but multiple fires.
- When setting these fires they tend to confine them to a particular geographical area, unless forced to move to relocate their activities.
- These areas of activity may occur within a one-mile radius of the fire setter's home or cover an entire region or state, even crossing state lines, with activity occurring all along an interstate highway. According to "Motive, Means, and Opportunity, A Guide to Fire Investigation" published by the American Re-Insurance Company's Claims Division in 1996; "Most serial arsonists walked to the scene of the fires they set, with 70 percent of the fires set within a radius of two miles or less from the residence of the serial arsonist." However large or small this area of activity is, it is referred to as the MFS's pattern (also known as geographic pattern).

- Most of these fire setters tend to graduate to bigger fires or escalate the level of violence, always looking for the better fire target, or "the big one". Some MFSs will continue to set the same type of fire throughout their career. In these cases the number of incidents will continue to increase with no known cause, except that of being suspicious.
- These offenders will avoid confronting someone during the fire setting act. In some cases the suspect will remain at the scene of the incident, usually out of view, in a concealed area or mixed in with the crowd. (Cases involving the suspects standing in front of a structure taunting someone and then throwing an incendiary device are not the MFS style.)
- During the fire setting act the subject will avoid personal confrontation and contact, especially early on in their criminal career.

Once the fire is set, certain MFSs may try to stand out by attempting to rescue victims or extinguish the fire. This, however, is the exception; in most of these cases the suspect will avoid the confrontation. These behaviors differentiate the MFS from the two other types of offenders we must discuss.

The two types of fire setters not associated with MFS tracking are *mass fire setters* and *spree arsonists*. The traits of both mass fire setters and spree arsonists were noted by Dr. David Icove and Philip Horbert in their 1990 study entitled "Arson Criminal Investigative Analysis" published by the NCAVC, Quantico Virginia. Mass fire setters differ from MFSs because they involve a single offender who sets three or more fires at the same location during one distinct time period. Mass fire setters will have specific targets and will usually act alone. Most mass fire setters are looking for confrontation, an example of which would be the scorned lover. The fire setter is the lover (or spouse) who feels wronged by their partner and then ignites fires involving the personal possessions of the target of their anger (spouse), such as the spouse's clothing, car, wallet, and documents, etc. All these fires would occur within minutes or hours of each other at the target victim's place of residence or employment. The subject (scorned lover/fire setter) tends to wait for the spouse to arrive, in order to confront them.

The other type of fire setters that should not be confused with our defined MFSs is the spree arsonists. Spree arsonists ignite fires at three or more separate locations with no emotional cooling-off period occurring between their activities, such as drunken individuals who begin igniting cars, dumpsters, and structures after a championship-sporting event, or during incidents of civil unrest. An example of this would be the fires, vandalism, and mayhem that occurred in the city of Chicago during the 1990s, after the Bull's basketball championships and the Bear's Super Bowl win. Spree arsonists will vary locations, picking targets randomly and may be acting in-groups such as mobs during a period of civil unrest. Incidents that occurred in the city of Los Angeles during the 1960's "Watt's Riots" and following the 1993 "Rodney King verdict" would be examples of spree arson.

Targets of a spree arsonist might be a car fire on one block, a grocery store fire two blocks away from the first, and a dumpster fire at the rear of the strip mall across the street from the grocery store. All these incidents would be occurring either simultaneously or within minutes of one another. However, spree arsonists should not be confused with mass fire setters who also ignite fires in the same general location. Mass fire setters have a great deal of emotional involvement in the target, whereas the spree fire setter has little or no emotional connection to the target.

If that was not confusing enough, ask: what is the main characteristic that will separate these two types of fire setters from the MFS? Time is the key factor in defining the difference. Remember, mass and spree fire setters have a timeframe factor. This differs from MFSs, who have a different relationship with time. MFSs always set fires according to *their* schedule and when it suits *them*. Regardless of the motive, they will ignite a fire when *they* feel like it. This need to set fires on their schedule will control the events leading up to the actual ignition, which is the result of a plan, as opposed to a sudden impulse or accident. MFSs choose the location, the ignition source, as well as the entrance and exit routes prior to setting the fire. (Note: This primarily refers to organized offenders. Disorganized offenders occasionally choose targets and times, while organized offenders are more likely to due so. These two types will be discussed later in this chapter.) MFSs believe that time is on their side and that they will be able to operate when it is convenient for them. It is this belief that will facilitate tracking them. Forensic psychologists refer to it as the emotional cooling-off period. For the purpose of our process, time, which is a very important characteristic, will be considered a physical feature. Our world is ordered around the physical use of time. We set our work schedule by time. Time is the measurement we use to record and determine sports records and then attempt to break them. Time is an important concept around which most economically developed societies function. In this process it is especially important, because we document and track patterns according to time.

Disorganized Vs Organized

There are approximately eight categories or motives describing fire setters that are most widely recognized. From the professional profiler's perspective there are two types of MFSs or offenders. For the purposes of this discussion about MFSs, we will focus on six categories, which we will compare with these two types of offenders. As explained by Gordon "Gus" Gary, a profiler and a supervising agent in charge of the Mobile, Alabama, ATF Office, profilers refer to two types of serial arsonists, the *organized* and the *disorganized* offender. Many of the characteristics displayed by both are directly related to the tracking process.

The disorganized offender picks random targets or targets of opportunity such as rubbish or used tires. They use available materials and set the fires with open flames using matches or cigarette lighters. The burns are unplanned and the severity of the burn is unanticipated. They usually have limited mobility that is frequently indicated by the location and type of fires. They often set fires close to their residence and tend be nocturnal in their habits. Disorganized offenders tend to be loners with few friends, who many times feel rejected. Finally, these offenders are very likely to be alcohol or substance abusers, as well as opportunists, unlike the organized offenders who are planners.

Organized offenders make preparations and use ignition devices to control the severity of burn. They plan the access to and escape from the crime scene. They tend to be mobile as indicated by their crime scenes and tend to live some distance from the crime scenes. These subjects tend to be egocentric with an indifference to society in general. They are manipulative, cunning, and methodical in their planning of the crime and this gives them feelings of control and power. Table 2-1 shows a comparison of the two types.

DISORGANIZED	Versus	ORGANIZED
Random targets		Selective targeting
Open flame		Devices used to ignite
Use of available material		Preparation to detail
Severity of burn unplanned		Planned ignition & burn severity
Limited mobility		Highly mobile
Loners		Egocentric
Feels rejected		Indifferent to society
Few friends		Manipulative, cunning, and methodical
Sets fires near home		Lives some distance from fire scene
Substance abuser		Chameleon personality

Table 2-1:
Comparison of Offender Types

The Six Categories of Multiple fire Setters

Vanity or excitement oriented

These fire setters may have personal goals they are attempting to achieve through fire setting. This will occur when suspects are unable to achieve their goals through the normal process. Examples of these individuals would be public safety officials (police or fire) or private sector employees (security guards) who attempt to gain some recognition, possibly by being first on the scene or through some heroic act; they anticipate being perceived as needed. Other examples would involve suspects attempting to gain employment.

This occurred in southern California in 1992 when two part-time firefighters who were seeking full-time employment with the local fire department set a wildland fire and attempted to extinguish it. They believed that their efforts would demonstrate to the fire chief how capable they were. Another motivation is to satisfy the need for excitement via the act of fire fighting because these individuals are action orientated. Unfortunately, the fire service job is predominately filled with mundane and tedious training, so when the job does not supply the type of adrenaline rush required by the subjects, they will seek other means to fulfill the desire for action and adventure. For some, that means doing what they know best—fighting fire. Although these MFSs can be either organized or disorganized, most will be organized, unless the offenders are using some type of substance abuse, in which case they will tend to fit more of the disorganized offender profile.

Anti-government, anarchist, or extremist groups

These subjects are motivated by what they perceive as some injustice done directly to them or others by any and all forms of government and institutions. They do not abide by the normal forms of political action or protest. They protest using violence, which is intended to bring attention to their cause. Examples would include extreme animal rights activists, fire bombers, echo-terrorists torching research facilities, extreme anti-abortion activists torching women's clinics and white supremacists groups attacking religious and government structures.

Of the six MFS categories, the offenders of this one are the most difficult to track and catch because they are among some of the most intelligent and most motivated individuals who set fires. Although most follow the predictable progression in fire setting, offenders in this group tend to choose secluded locations to manufacture and test their incendiary skills. Only when they are effective at their craft do they make their presence known. These offenders are even more difficult to track because of the rural communities where they prefer to set up shop. In suburban and urban communities the testing of incendiary devices attracts attention, whereas in more remote locations, an incendiary could go completely unnoticed. Another problem is that these MFSs will operate over multiple jurisdictions and across state lines.

Do not give up hope of identifying and tracking these MFSs however, since technology and federal help are on the way. As for the latter, the introduction of federal laws to combat domestic terrorism will bring with it the assistance and full weight of federal law enforcement agencies along with the prosecutorial community. This occurred in the mid-1990s with the Federal Church Fire Task Force, which investigated fires throughout the US. Furthermore, a significant aspect favoring the tracking of these MFSs is the fact that these unsubs seek to bring attention to their cause and by doing so, bring attention to not only their cause, but to themselves as well. For now, concentrate on the time factor, for anti-government MFSs will tend to follow the organized offender trend, where historical dates and historical sites frequently play a significant roll in the fire setter's activity. Anarchists will combine forces with them looking for targets of opportunity within larger events. This observation was noted at the 1999 World Trade Organization (WTO) conference held in Seattle Washington, which was targeted by anarchist as well as labor, anti-labor, and environmental activist. The riots that followed caused the city of Seattle to re-evaluate all their public events in the future. The key will always be to identify and anticipate when an event may be the target of this type of MFS.

Spite/revenge

These suspects have been, or feel as though they have been, wronged, mistreated, or discriminated against. Because they do not wish to confront those who have wronged them, the suspects use fire as the destructive tool of their revenge. These fire setters target subjects or facilities, which they feel, represent those who have mistreated them. They include employees passed over for promotion or dismissed from their post, or individuals who feel they were insufficiently reimbursed for services rendered, individuals who have been bullied by other individuals or groups or suspects who have been refused services at stores, restaurants, or bars. These offenders will not only set fires at targets of their anger, but also call in malicious false alarms (MFA) to those locations, possibly using names of subjects against whom they seek revenge. The spite/revenge fire setters can fall within both offender types but tend to more closely resemble the organized offender. In most cases suspects can be quickly identified, unless multiple subjects feel they have been wronged, as in the case of union employees on strike or the relocation of a business or industry out of a community or region. The use of a calendar to track fires can be very useful during this type of investigation. It is possible to note direct relationships between the announcement of meetings, issues, and fire activity.

Malicious mischief

This type of fire setter is exactly what the term implies. As defined by David Icove and Phillip Keith in their 1983 paper "Principles of Incendiary Crime Analysis: the Arson Pattern Recognition System (APRS) Approach Arson Information Management" published by the Kentucky Department of Public Safety, malicious mischief is the needless destruction of property by subjects using fire as their weapon. This is sometimes referred to as vandalism, especially when committed by young adults and juveniles. They should not be confused with the juvenile fire setter because the subjects are not exclusively juveniles and young adults, although they seem to be the most common suspects. The malicious mischief types of MFSs usually leave an excellent trail that can be tracked. Since these suspects have a great deal of free time, they tend to leave themselves open to an established pattern and usually act in groups. Their ability to cause destruction and escalate the level of violence should not be underestimated. As with all MFSs, these offenders will start small. In some cases when only one individual is involved, the subject may be satisfied with causing only minor mischief. However, when groups of individuals are acting in concert, minor incidents will not satisfy their thirst for violence. Malicious fires could be the group's initial acts of violence, thus transforming a group of individuals from an association into a criminal gang. Malicious mischief fire setters tend to be disorganized offenders. In Philadelphia, 1996, a series of fires occurred that were dubbed the "Badlands Fires". A group of five young males between the ages of 13 and 19 began setting fires to earn points as part of a game. The object of the game was to see who could achieve the highest score. Points were awarded for the most misery caused by setting fires and for eluding the police. Over time, these thrill-seeking youths progressed to setting fires that destroyed occupied structures. The suspects were eventually caught by the police and fire marshals, but not before setting some 30 fires. This case was documented very well in a television series called "Blaze" produced by The Learning Channel in 1999. The suspects' largest blaze was the I-95 fire, which burned for three days and shut down the main interstate through the city of Philadelphia.

Mentally defective

It is this group that we envision when the terms *pathological fire setter* or *serial arsonist* are used. This category contains many subgroups including *compulsive fire setter* (pyromaniac), and *psychopath*. Serial arsonists are defined as a person or persons who are involved in three or more separate fire-setting episodes, with a characteristic cooling-off period between fires that may last days, weeks or even years, with each additional episode escalating the level of violence. Most MFSs fall under the umbrella of this definition. The mentally defective MFSs are the most dangerous and difficult to identify, due to the unique selection of victims and unpredictable gaps between incidents. Compulsive fire setters (once known as pyromaniacs) are defined as psychotics or borderline psychotics, who feel the need to set fires all the time. Sometimes referred to as "torched souls", their pattern of behavior is the result of a personal tragedy or some type of neglect. The origins of their compulsive desires or urges to ignite fires have frequently been debated within the psychiatric community. Some studies on biological causes for the pyromania have caused some to theorize that a metabolic or neuro-transmitted abnormality may cause the desire to set fires. Currently, this has not been proven. The results of one study of fire setters as explained in the article "Pathological Fire-setting 1951-1991: A Review" by W. Barnett and M. Spitzer appearing in the 1994 Medical Science Law Journal suggested personality disorders. The subjects studied could not deal with stressful situations, disappointment, or insult. In many cases their personalities were characterized as having an "intolerance of frustration, over-inhibition of aggression in ordinary behavior, striving for power, readily feeling insulted, and a tendency to act out aggressively". It should be noted that the studies are still being conducted to support these findings.

No one really knows why the compulsive fire setter ignites these fires, but what is known is that they can't set just one. The mentally defective also include the psychopaths. Fire psychopaths are usually associated with sexual behavior for stimulus to set fires. Although this is a possible motive, it is not the only one for psychopaths. Some of the fire incidents involving this type of fire setters show signs of ritualistic acts. The offender is angry and/or power driven. Some psychopaths who are fire setters will show traits of the "homicidal triad" that includes cruelty to animals, bedwetting, and multiple fire setting. This was true of David Berkowitz, commonly known as the "Son of Sam". One fact that is confirmed by most studies of mentally defective fire setters is that women are in the minority; most MFSs are male. Mentally defective fire setters will vary with the individual in reference to organized and disorganized offenders. For example, psychopaths will have a tendency to be organized offenders, as opposed to compulsive fire setters who tend to be disorganized offenders.

Juvenile fire setters

The FBI crime index identifies juveniles as being responsible for over 40% of all incendiary fires from 1986 to 1993 and responsible for 50% or more of those from 1994 to 1998. Although juveniles are usually some of the easiest fire setters to catch, the motive is not always as easy to identify. The subject of juvenile fire setters is a long and complicated one. Juvenile motives for setting fires will vary with the subject. For this reason it would be difficult to track all the different types of juvenile fire setters. These subgroups are explained in the work done in 1986 by G.A. Sakheim and E. Osborn called "A Psychological Profile of Juvenile Fire Setters in Residential Treatment: A Replication Study" in Child Welfare. The subgroups, from a psychodynamic aspect, are the curiosity seeking or accidental fire setter, the cry for help, the attention seeker, the would-be hero, the seeker of excitement or sexual pleasure and the psychotic. However, most are simply due to curiosity on the part of the fire setter. Only small percentages are the result of mischief, compulsive, or psychotic behavior. This does not mean that we should not track the activity of juveniles. As noted in the article "Juvenile Fire-Setters" written by David Liscio, published in Firehouse Magazine, Sept.

1999, of children who deliberately start fires 81% will become repeat offenders if not treated. The majority of these juvenile fire setters are disorganized offenders.

Arson for profit

Incendiary fire setting for the purposes of insurance fraud and economic gain would be almost impossible to track under the methods covered in this book. For that reason, it has not been tracked with this process. Suffice to say that in most of these cases, if the fraud were successful, the fire would be considered non-suspicious and tracking would be unnecessary. However, arson for profit does occasionally occur and can be proven. On a personal level I have only been involved in fire fraud cases where organized crime figures and/or individuals were paid to set specific fires among their many other criminal activities. There was one individual who, it could be said, worked part-time as a professional torch. This MFS who fit this profile closely, set two boat yard fires. His idea was to set a structure fire somewhere else in the community and then set the boat yard on fire hoping to catch the fire dept. off guard. The only reason he used the same MO (modus operandi) at both incidents was due to the fact that the first boat yard the offender torched was the wrong one, so he had to go back and torch the correct one a few days later.

Having said that let me add that I have never encountered an individual who made their living solely by setting fires. If they do exist, these types of individuals no doubt work for someone like a character called "Kaiser Sousae" from the 1994 movie, *The Usual Suspects,* or some presently unknown organization deeply rooted in more evil than I have ever encountered. As for the fire setter motivated by economic gain, it would be better to classify this offender under the "vanity" category of MFSs. This is based on the consideration that the offender has been unable to achieve economic independence or personal self-worth through a standard career path and will resort to criminal activity, constantly escalating the violence until the perceived goal is achieved. This individual would likely be an organized offender. We would theorize that profit driven MFSs are organized offenders, but since most of the information regarding motives is gleaned from subjects who have been arrested and prosecuted, it does not include the ones who managed to elude authorities.

Crime concealment

Due to the fact that this act of fire setting is secondary to the primary offense (*i.e.*, covering up a robbery or murder), it will not be discussed. Regarding fires set as crime concealment, if the unsub is committing multiple burglaries prior to multiple fire incidents, this would make the unknown subject a serial burglar as well. Therefore, it may be assumed that the unsub is also being tracked as a robbery suspect. If not, it would be prudent to track the unsub, but more information regarding the MO and signature details would need to be identified by theft experts, since the fire is only a method (usually unsuccessful) to cover the crime. This book will not further discuss this type of MFS.

Now that we have a very basic understanding of MFS types, let's move on and discuss how we track them. Keep in mind that these categories have been highly simplified and that it is completely possible that more than one motive may be at work with a MFS. Therefore, the assistance of certified experts is strongly recommended. For further information on motivational and/or psychological topics, please be sure to read the appendices of this book and reference the bibliography for the complete books, articles, and lectures on these topics. It is also recommended that you check with universities in your area for courses and seminars on these topics. The Internet can also be another good source of possible information. However, use caution before proceeding to discuss any of these topics. Be sure to check the legitimacy, reliability, and accuracy of each Internet site.

Chapter 3

HOW TO GET STARTED

Getting started in anything can be tedious and this type of work is no different. In the beginning, the process of collecting, sorting, entering, analyzing, and finally reporting will take time before you become efficient. Fortunately, once you establish the routine, it will take approximately 30 minutes (with a computer) to update and analyze the information from the previous day's activity. Gathering data, entering the information, and then sorting through it all can seem overwhelming. If that were not enough, the idea of starting this process can also meet with some resistance. Fellow investigators and supervisors will not be willing to spend the time or money on an unproven system. They may not see the value in or the return on, the effort spent to create this process. After all, why try to prove what we will find out eventually. Some fellow employees may feel threatened or resent this process, believing that this may affect their job security. Investigators may see it as a statistical record keeping system or a score card on which to judge their performance. Fear of the unknown is a tremendous factor to overcome in any situation. This process is no different. These problems have no easy solutions, but must be addressed if you are going to get the process off the ground. To deal with these problems, first remember that the key is to emphasize to the doubters and unbelievers what was covered in the first chapter. Continue to discuss the relationship between fire setting and more violent crime. Remind all parties that the NCAVC has noted a definite escalation in the number of serial arson cases.

If necessary, when starting out, pick and choose the cases with the highest value of return. These cases may not always be the biggest of cases, but they will be quick to develop and help prove the value. Call them targets of opportunity. For the law enforcement officer, an example of this idea would be to pick a high crime area with gang and drug activity. When gangs are involved with drug sales, fires are usually not too far away. Certain gangs tend to rely on particular weapons when carrying out acts of violence. We have noted that Central American gang members prefer knives and fire as their weapons of choice. Many gangs will use fire to intimidate witnesses and rival gang family members. They will also use fire to destroy hideouts and drug distribution and manufacturing locations. Work with anti-gang and drug enforcement units to help confirm these traits. These types of fires do not show the pattern search process well, but when you plot the fires and then the gang and drug activity on the map, your superiors will quickly see the value in what you are doing. If you are really fortunate and work in a community with little or no gang activity, you can simply plot standard crime activity using the standard community maps or computer mapping software. How you generate these maps and who you give them to will be covered in future chapters. For now, understand that it can be done relatively easily with minimal investment.

For fire officers, an example of this is malicious false alarms (MFA) or crank calls to fire and police communication centers. These calls can now be easily traced with enhanced 911 systems and caller identification. This can led to fast arrest if the suspects are adolescents. These MFA calls can be easily plotted on maps to show particular school boundaries or youth groups. Calendars can be created to show particular days of the week and month when the incidents occur. Use the plotted map and calendars to see if the information coincides with some function or activity such as early release, holidays, or school plays. Show the information to school administration, youth probation officers, and supervisors of youth programs to see if they note any similarities with scheduled activities. Finally, check with the callers who reported the incident and/or the location where the calls were made (such as pay phones) to see if anyone knows about the calls or who used the phone.

Based on the fact that most adolescents usually act in groups or tend to brag to other associates about their actions, it is possible to gain further information quickly by confronting the suspect with the information known. Then the suspect can be confronted with specifics about the incidents. Ask questions such as "Why set the fires on a particular day or certain nights?" and "Whose idea was it to set fires every other week?" Then there is my personal favorite: While discussing the fires with the suspect(s) hold a videocassette tape and tell the suspects that it is from the surveillance cameras located in the 24-hour convenience store near the location from where all the false alarms were phoned in. Let the subject infer the rest. It is possible that they will believe that someone has informed on them. For juveniles it is usually no fun unless you can tell someone. Furthermore, some adolescents will give up information quickly, especially if they believe the secret is out. This simple ploy will usually gain good results with admittance of guilt from one or more of the parties involved.

If sometime during the process a skeptic asks why you can't just go to the location where the calls were placed, ask questions, and forget the rest, explain to them that the reason is that the parties involved (including the suspect) will know that all you have is the location of the call. Instead of being in the position to know how many calls, when the calls were made, where all the calls originated, who was called, and why calls were placed on those particular days, keep the suspects wondering who gave up all that information.

For the fire personnel, this process can also be used to prove problems with faulty or chronic automatic fire alarm (AFA) systems. Many times, chronic AFA problems will not be noted or identified due to the same reasons why MFSMFSs are not identified. Multiple shifts working different hours with varying personnel (due to vacation and leave) make it difficult to keep up with daily routines, not to mention unique problems. AFA can be easily identified and plotted to show how services are being used and deployed. Your superiors will then have documented evidence to show elected leaders and lawmakers that there are issues that need to be corrected.

Another good argument for the use of tracking is economic. If a community discovers that they are the victims of a MFS, there may be some economic recourse. For example, in 1991, when the Cities of Ocala and Gainesville, FL, identified themselves as the targets of MFSs, the two cities looked toward the federal government not only for law enforcement assistance, but also for reimbursement. The federal government sent the two cities funds for expenses related to the suppression, the investigation, and the public education effort. Under an emergency federal law enforcement grant, the two cities were able to receive funds that they would have otherwise been responsible for. By identifying the MFS problems (that the cities were the targets of church fires), the communities were not burdened with that expense. If you are still encountering resistance and the question continues as to why tracking is an important tool, remind them about the "Son of Sam".

To get started, you will need a shopping list. If possible try to get some computer equipment (hardware and software). This is not always possible, but if it is, refer to chapter 11 where a list of general requirements has been laid out. The tracking process only requires a stand-alone desktop or personal computer, the type sold in most department stores. If you have not had any luck getting equipment from your agency, do not get discouraged; there are other sources. For one thing, state-of-the-art equipment is not required to accomplish this process. Remember that this process can be accomplished by hand. However, new or state-of-the-art equipment will make everything but the collecting of information go much faster. Even so, if state of the art is not what is being offered, do not worry; you can get by with someone else's seconds. Computer experts would say I was working in the dark ages if they observed the age of the equipment we started with. Just to give you an idea of how old our computer equipment was, when I started with my agency, we were excited to have a computer that would take 3.5" diskettes and had a color screen.

How do you obtain this equipment? First, begin by looking at sources for funding equipment. Some recommended sources are through federal and state grants. This is how the Fire Marshal's department of the State of Maryland Fire Marshal got started with their programs. Using federal grants, they developed a statewide data sharing system. Maryland also used federal funds to develop a management guide for the fire investigation unit. The first program, called Maryland Fire Investigation Data Reporting System (MFIDRS), was developed in 1996 and is overseen be the State Fire Marshal's Office. The second program was a group effort overseen by the Maryland Fire and Rescue Institute.

This type of federal and state grant usually requires proposals, expense breakdowns, agency statistics, and community information. In some cases, grants may be for specific amounts, which are more than you had considered. In these situations, see if you can include other departments or agencies that would benefit from the grant. Always try to be open to all types of grants. Remember that computers can perform more than one task. A computer purchased under grants to keep track of burn reports can also be used to keep track of fire incidents. Simply consult your local computer expert to make sure the equipment can do both before purchasing.

The second recommended source for funding of equipment is the private sector. These sources would include, but are not limited to, the insurance industry. The insurance industry has always been a good source of funds for fire investigators. Some insurance companies directly fund law enforcement programs; especially those related to fire and arson investigation. Suffolk County (where I am employed) has received grants for computers, arson rewards, and even children's T-shirts for our canine unit. Examples of companies who could possibly help would include the Factory Mutual Company that sponsors fire investigative programs; Aetna Insurance that sponsors arson prevention programs; and State Farm and Safeco Insurance that have both sponsored arson canine programs. Large corporations in your area would also be a good source to check with. Some examples would include defense contractors, the health care industry, computer manufacturers, computer software companies, and the shipping industry. In many situations, these corporations want to develop programs that place funds back into the communities where they reside. Check with various levels of your company to see if anyone on staff has written grants for federal and private funding. These people will be a great asset, not only for writing the grants, but also for leading to other sources that were not thought of originally.

What if you have not had any luck in the grant funding area? Do not be discouraged. Based on the fact that state-of-the-art equipment is not needed, other sources may be available for hardware. If you have any federal facilities in your jurisdiction, check with them. My own agency has had great success finding equipment that has worked for us through the lean times. Educational institutions and large corporations are also good sources for hardware; they are constantly upgrading systems. Some of the sources mentioned may tell you that they cannot give the equipment to you. The reason for this is based on the fact that by your requesting the equipment, it can be interpreted that the equipment has some value, therefore, the agency should be attempting to sell the equipment. For that reason, it is easier for the agency to junk the equipment. Rather than attempt to sell the equipment, which has little or no resale value and would require more paper work and man-hours than the sale could generate, it is easier to just discard it. If this is the case then say "no problem" and then ask if they know what day they will be discarding the equipment. Be sure to be there on that day and tell them you will be happy to save them the expenses related to disposing of the equipment. Along this same line of thought, do not forget to consider other departments within the government, especially those who do data processing such as the Geographic Information System (GIS) or Management Information Systems (MIS) people. These are the payroll and records storage people. They tend to upgrade and update their equipment regularly. Remind them that you are only looking for one desktop computer and not some mainframe system used to run the whole municipality.

In regard to software, once again you can try all the sources mentioned earlier. If you do decide to purchase new software, be sure to verify that it will work with the computers that you are using. If the computers acquired are meeting the bare minimum requirements of the software, then it is recommended to look for something more compatible with your machine, possibly utilizing older software. Older software can be found in the markdown rack at the computer store and at computer shows. There is an advantage to using discontinued old software that most users do not realize. This is based on the fact that all the bugs and glitches in the systems have been discovered. Computer experts will say that you do not want that because it will not do this or that and it will give you a problem with something else. With state of the art, they may have fixed the known problems, but are unsure of future unknown problems. With the older discounted software you will know the limitations and, as long as it does not affect the tracking process, you can just avoid performing any task that is known to create problems. In other words, work within the limitations of the software.

In your search for software, don't forget to check with the same sources within your own local government mentioned earlier where you inquired for hardware. All software purchased by governments is licensed to them. Your agency is part of that government, which means that you should be able to gain access to it. Once again, agencies within your government that have been using computers for many years, may have older software sitting on the shelves collecting dust. This software will usually work well with that donated non state-of-the-art computer hardware. Furthermore, because the older software would have been used by your own people, they can help you learn and set up your system.

If you have had no success in attaining any computer equipment, the tracking process can still be accomplished with a few hours of hard work at the local public library and some paper sorting in the office. Simply get all the information for data entry prepared at the office, contact the public library (or local college), and ask to sign up for some computer time. Also, see if they have some basic database software available for use. They may even have videos available for software training. Give yourself a few hours and make sure someone is available to take you through the basics of the programs. Request time that coincides with your shift and get to work.

Try to attempt at least one or two of the ideas from those listed above before proceeding by hand. If you wish to proceed with conducting this process by hand, sit tight and continue reading; all the requirements necessary will be covered in chapter 6. On the other hand, if you are starting to think that these ideas seem to be too much work, do yourself a favor and try to return this book or pawn it off on someone else in your office. If you are not motivated enough to follow through with one of these ideas, you will certainly not be motivated to complete this process.

In this chapter we have discussed the key items in getting started. They are not the only problems or issues you will be confronted with. Remember that in some instances you will have to be unconventional in attaining the equipment. This does not mean that you have to steal it. What it does mean is that you need to start thinking outside of the box. If you need to use forfeit/seizure asset to purchase items, do so. This may mean sharing computer time with some other squad, bureau, or department within your government, but as you become involved with other groups, opportunities to obtain your own equipment and software will begin to arise. Unfortunately, if you have not interacted with other groups, the opportunity may never arise. You may even be compelled to spend your own hard-earned money to help accomplish the process. When I began my tracking and mapping system, I had to purchase the thirty-five dollar computer mapping software that the process relied on. It was some of the best money I ever spent. To this day, my supervisor and I still prefer that software over the thousand-dollar GIS software we currently use for our county-wide geographic mapping data.

Chapter 4

WHERE TO GET
THE INFORMATION

At this point it is important to discuss the next phase in this process, determining the information needed and how to gather it. The information sources required will vary, but can be broken down into three categories. For two of these categories we will borrow terms used by the espionage services of the world. The first of these terms is referred to as HI (human intelligence) and EI (electronic intelligence). HI would be the obvious assets that investigators have been using for centuries and would include sources such as a CI (confidential informant). If you have little or no background in this area, we will cover some basic thoughts about informants. The first basic rule is that most informants will only tell you what will benefit them and tend to forget or leave out what does not benefit them. When dealing with CIs, try to limit the details and not give up too much information. This is based on the idea that informants are always re-evaluating their situation and position with you. If they see the opportunity to tailor the information into what you want to hear, they may do so. Remember, most of these people did not become informants out of the goodness of their hearts. These informants will have to be evaluated on a case by case basis; use your own judgment. If you are not an investigator or police officer and feel that this would be a difficult relationship to cultivate, remember that informants are only one human source for information. Do not limit your HI to these sources alone.

HI should also include those professionals who have related public safety functions. Primarily, this group will include fire service personnel such as firefighters, dispatchers, and fire inspectors. These individuals will not only know about activity in their own community, but also neighboring ones. Fire prevention officers will also know about probable targets and may be able to identify probable suspects. Also included as HI would be building inspectors, eviction officers, and, of course, police officers who patrol locally. If possible, probation or parole officers should also be consulted later in the investigation for ideas on probable suspects. Unlike CIs, these sources are professionals and will have little, if any, reason for holding back information. Keep in mind that one of the rules that applies for CIs also applies with these sources; namely they will tell you what you want to hear. Try not to direct your questions too closely to the exact information you are exactly looking for. Be general and take the good information along with the bad. Allow them to be neutral and unbiased in their answers. This can be very helpful later on in the process if a pattern is discovered (based on facts surrounding the incidents) and confirmed by unbiased professionals. As with your CIs, continue to judge the information on a case by case basis. Rely on relationships that have been built over time.

The next source for information to consider is the general public. This group would also include residents of the neighborhood and service professionals such as barbers, convenience store managers, or local café and restaurant owners. Also included in this group would be community groups such as churches and rotary clubs and civic leaders. The general public resource group can be the most difficult to question. Issues should be kept as vague and general as possible. If the individual wishes to discuss specifics, let them continue, but always try to return to the general issue once they have told you all they want to tell you. When dealing with the general public, remember they are not trained professionals or CIs. Treat them with respect, but remember they also may only tell you what they want you to know and more than likely, they will only have a portion of the whole story to tell. I have never turned information down even if it was not useful. You should always make a point of thanking these sources and letting them know that their help is appreciated. You never know if some day in the future that person may have more information that will be helpful or in those rarest of cases the person may even turn out to be your suspect.

Another source for HI is eyewitnesses. Eyewitnesses should always be handled as separate resources with information on a specific case. Eyewitnesses should always be directed to the lead investigator of that particular case (if not the person conducting the pattern search).

One of the least effective sources (in my opinion), but nevertheless a source of HI, is the local news reporter. At this point I would advise that you "Proceed at your own risk" when using a news reporter as a source. The success or non-success of this will be determined by your relationship with the news media in general. Keep these simple facts in mind. If the reporters have information that you do not, one of three things will occur. First, if they respect you and/or your office, they will call before going with the information in their story. The call will be for comment (this usually means they have some respect, but are also fishing to see how good their information really is). Second, they are calling to let you know this is what information they have and when they are going to air it (this usually means they have good respect for you and/or your office). What they are really saying is, "I will show you mine if you show me yours." In this scenario, some kind of deal can be worked out to get the reporter's information and, in return, the reporter is given the exclusive to the story. The third scenario would be where the media have little or no respect for you and/or your office. They will try any way they can to gain information from you, commonly known as a fishing expedition. That could mean telling you they have information related to a case when in reality they have nothing and hope that you will give them something to work with.

The last form of HI to remember is you. The sighting of smoke on the horizon near the area in question while on patrol can pay off better than any other source. If possible, observe activity in the fire area. Take note of the terrain and/or conditions of structures in the fire area. This, coupled with many of the issues and ideas covered in this book, will help you to become an observant HI source.

The second term that we borrow from the world of espionage is EI (electronic intelligence). EI comes in various forms, all based on observation and monitoring equipment such as store cameras and burglar and fire alarm systems, to mention a few. EI systems have been around for years and their capabilities have improved at amazing rates. In chapter 13 we will touch on future observations systems, but in some areas the future has arrived.

For instance, in parts of the Northeast and the West Coast, a network of closed circuit cameras observes major interstates and highways. These cameras have good resolution and can view locations and events in what is called "real time". This term means that the images as close to live as possible with no delay for developing or reproduction. These cameras used are perched above the roadway with a zoom function and the capability, in most cases, to move in a 180 to 360 degree radius. If the fire activity is occurring in the same area as the cameras and during off-peak traffic hours, it is completely possible that the road monitoring center of the state's highway department may be one of the best places to set up surveillance. These same type of cameras are being installed at parking garages, schools, government facilities, and even in news media centers. These are only a few examples of EI; there are many more to research.

Personally, I have had the most success with monitoring fire band radio frequencies. For a short time in my fire service career, I worked as a fire communications dispatcher. While working there, the gentleman that trained me was a seasoned twenty-year veteran. One day during my training a neighboring department signed on with a brush fire and my trainer commented that there was something very odd about that department. He went on to explain that it seemed as though every time this particular fire department's specially designed off-road fire truck was in service, they had grass or wild land fires. He continued to explain that, when that specialized truck was out of service (for whatever reason) there were no grass fires in that community. Later that year, I noticed how small the fire district was in that community. I also noticed how little grass or wildland there was in that community. From that point on I began paying attention to radio communications and noting what was going on around the communities where I worked. Unfortunately no one else on our staff had considered monitoring the fire radio traffic as this type of resource. It was during my off hours when one suspect did most of his work, ultimately killing. If we had realized the full potential earlier and been paying attention things may have ended differently. To this day, I still do this with a scanner monitoring the fire frequencies. It has paid off, allowing for a rapid response to areas we were concerned with. This type of response will allow the investigator to gather information and note items that cannot be seen from reading reports or looking at maps.

The Connecticut and Massachusetts State Police have used this technique for multiple suspicious fire incidents. An accelerant detection canine unit is assigned to lay in wait in the neighboring community and respond with fire personnel at the first notification of fire. Used in the 1990s, it was refereed to as rapid deployment. In the 1980s a similar concept called "Red Caps" was employed in New York City and neighboring counties. It involved federal funding to allow additional fire investigators to be working during hours of peak fire activity. This concept allowed investigators to respond on the initial fire alarm activation. The success of this program varied, but unfortunately ended when funding stopped. The final point here is that HI (you) must interact with EI. In the best of worlds, this will result in the best results. But what if you do not want to listen to that chatter all day? Then you will have to rely on other information resources.

The third category from which most information will be gathered is another combination of HI and EI sources. This will come in the form of documentation. The sources for this documentation will include fire department communication logbooks, run sheets, and daily statistics. The other source will be police logs, blotters, and field reports. The location of these documents will vary with agencies. Depending on the jurisdiction, this information will be in a computer log, bond journal, or written log. A good place to start looking with regard to the fire service is fire communication centers. Try regional centers first and then local dispatch centers. There is no set document style, so once again they will vary from agency to agency. Some fire departments may use nationally recognized forms such as the standard form #904 designed by the NFPA (National Fire Protection Association). The NFPA states that their statistics capture about one third of all US fires each year. The NFPA also states that more than one third of all fire departments in the US provide data.

Additional sources include NFIRS (National Fire Incident Reporting System). This system was started in the 1970s by the United States Fire Administration and has been handed off to all states. NFIRS data has been used for public education, building codes, and consumer product safety. The system is voluntary, but widely used by the fire service. Private vendors such as the Fire Service Software Vendors Association who have created software to meet standard requirements and also tailored the system to meet special needs of individual states and municipalities have also helped to improve the NFIRS system. According to the article, "1999 National Update" by Jay K. Bradish in Firehouse Magazine, Sept., 1999, only two states were not participating in NFIRS or a similar data recording system.

Other data resources are police department blotters and field reports. The FBI has been collecting crime data since 1930. The DOJ (Department of Justice) has required UCR (the FBI's Uniform Crime Reporting System) reporting in order to apply for some federal grants and funding. Similar records keeping systems exist at the national level for law enforcement as with the fire service. Both the UCR and the NIBRS (National Incident Based Reporting System) are voluntary, but are well supported by most law enforcement agencies. Whatever type of information source system is employed, be sure to identify that source in the final report. This will be important for verification and further research by other investigators. This is what is known as *meta data*, that is, the identification of the primary information sources from which all other databases are developed. Meta data has become a much more important issue in the computer age because so much data can be processed in a greater quantity than ever before. This greatly increases the potential for inaccurate and/or outdated information to be used when formulating databases. Therefore, if the information cannot be verified, do not use it.

Although all of these systems refer to themselves as uniform or national, they are neither. As mentioned earlier, no agency is required to report. Compounding the problem is the fact that formats between the different systems will vary. This may require the greatest effort if the information is not well defined within the documentation. It will also require you to be more flexible in your gathering of information and in reporting the data. Remember how the term "set or ignited fires" was defined. Although it may not be against the law to intentionally ignite a pile of old newspapers located at the street curb, for the process of tracking, we want to know if it happens. For that reason, it is strongly recommended that you have at least two separate documenting resources when gathering information. Never rely on one documenting source. Try to identify how all the different types of incidents that will be used for tracking purposes are categorized. For example, in a police documentation resource, wildland fires may be identified in the same manner as a structure fire if a 10-code communication system is used (codes such as 10-4 for affirmative or 10-1 and 10-13 commonly used for officer needs help). The same is true of fire logs that may list an accidentally activated structural fire alarm and an incendiary structure fire as the same event. You will also need to check to see if AFA or A/A (Automatic Fire Alarms) and GIC (Good Intent Calls) are differentiated from MFA (Malicious False Alarms). To further complicate things, agencies tend not to define the term "suspicious" in the same way. Thus the use of this term by the local fire service may vary from that of the police department. The pile of old newspapers burning at the curb at 3:00a.m. may not be viewed as suspicious by the police department, but at the same time may be viewed suspicious by the fire department.

Do not become discouraged; remember, this is a learning process. Once you learn the basics, you will be able to differentiate between the varying quirks or uniqueness of each documenting system in the future. If an agency says that they do not keep documents such as the type you are looking for, try another office within that agency. Certain individuals within the agency may not understand what it is that you are looking for, while others may not be sure if the information can be released to you. Though access to the information should not be difficult, it may still require a letter or an official request. Most information is public record and will not infringe on any individual rights. Furthermore, most agencies now keep documentation for annual reports and job security. Public safety agencies need to show what citizens get for the tax money being spent and why these agencies require more in the future. Regardless of the format, various reports should contain most or all of the information required for the tracking process. If personnel still hesitate to supply information, remind them that you do not require names or phone numbers, only time, date, type, and location. Consider it a perk if the agencies will also supply the name(s) and phone number(s) of the person(s) you are inquiring about.

As to probable cause for arrest, the tracking process is, to my knowledge, not scientifically proven or accepted in court as a scientific process. (This would be defined by the case names of Fyre and/or Daubert. You may want to check to see which legal definition is used in your jurisdiction.) Instead of seeking to use the process as part of the "evidence", think of it as being no different than the one used by that deputy marshal we mentioned in chapter 1. He used the surrounding environment and the general public to track down his man or woman. Accordingly, the tracking process can also be compared to the one used by hunters who track their prey through the snow, over matted grass, or past broken branches. Because of their persistence they always get their prey. That is how the fire tracking process is done; by establishing how the suspect has a link to the environment based on the pattern searches. This link gives you reason to talk to the general public in and around the active fire area as well as probable cause to stop and talk to subjects in the same area. Future chapters will discuss what steps to take when most of the information received is based on a tip or hearsay with little else to go on.

Chapter 5

WHAT INFORMATION IS REQUIRED?

In chapter 2 we discussed the questions of who set multiple fires and why. In the last chapter we discussed the question of where to get information on persons who may have set fires. In this chapter we will complete the list of the basic investigator questions of what, where, and when. It is interesting that the issues concerning where and when for this process are also two of the most important issues for business and military leaders. One of the first concerns that leaders of industry note is that "time is money". Military scholars and leaders say, "time is of the essence" and when planning an attack "timing is everything". For the tracking process, the question of when is also a question of time. That time (or when) not only refers to the time of the incident, but also to the time period when the meta data was collected and from where it originated.

To know these things, we must define time. Time is not a physical feature. It is a man-made idea that we base our entire social order on. To maintain this social order, time has been changed into a physical feature similar to a place, setting, or location, where degrees are subdivided by hours, minutes, and seconds. Before the advances of "Loran" navigation equipment and GPS (Global Positioning Systems), the technology of the day dictated that ship's positions (specifically longitude) be calculated based on time clocks. The accuracy of these timepieces would dictate the ships navigational ability. Then mankind developed atomic clocks used to keep the most accurate form of time. This is more efficient than determining time by the rising and setting of the sun since in certain parts of the world the sun does not set for months and then does not rise for a few more. Time, as we know it, is based on a system where seconds and minutes are in intervals of sixty. For all practical purposes, this system works relatively well. Therefore, let us use it as all other institutions do, to our advantage.

We will use the keeping of time and the documentation of it to base our pattern searches. The tracking process requires the basic information and the time of the alarm or the time the incident was reported. This recording time will rarely be exact. In most cases, it is an approximation of the report time, usually within plus or minus five minutes of receiving the notification. This is usually the best that can be accomplished and will satisfy the requirements for the tracking process. Do not confuse the time of the report or the notification of fire with the time that the fire was actually initiated. That will also be an approximation. The majority of fires are recognized within 60 minutes of ignition, but the exact ignition time is very difficult to identify. Depending on the location, the actual ignition could have occurred minutes, hours, or even days before the fire is discovered. However, to have any success at tracking the time of ignition, you will have to determine the actual fire ignition to within 60 minutes. Ideally, the notification time will occur within 12 to 24 hours of the ignition. This allows for the actual ignition time to be calculated as accurately as possible. Of course, all of this requires that the point of origin is identified and then the time of ignition can calculated. When it is determined that the time between ignition and notification is greater than 60 minutes, a fire investigation will be required. Obviously, the closer to the exact time of ignition, the more accurate the tracking report.

For those communities where the majority of recognition occurs 60 minutes after ignition, the following issues should be considered. First, delays in recognition in rural settings are usually based on access. In most rural communities access and the means of travel will be limited. Therefore, although it will take an hour to recognize the fire, it could take 30 minutes to 60 minutes for the MFS to travel to the fire and then depart. By calculating this fact on the maps (covered in chapter 8) a patrol area and/or surveillance can be established within the 60 minutes before and the 60 minutes after the incidents. When multiple incidents have occurred in this setting, the travel distance can be used to successfully draw the net around the unsub.

Secondly, in urban settings, fires may be ignited in large abandoned structures that are a city block or more in size and two or more stories in height. Many of these structures are of heavy timber or fire proof design. These structures are old abandoned buildings with outdated production equipment and supplies remaining in place. Examples include abandoned machine shops, defunct heavy industry, and old lace or textile mills. These types of buildings will appeal to the MFS due to the vast size of the structure, which enables the offender to set the fire deep within the bowels of the structure. In these cases, location, and type of incident will play a larger role that we will discuss later on in this chapter. Knowing that this type of fire occurs in abandoned structures, these targets can be singled out from other types of structures. Structures that meet the unsub's MO can be identified, searched, patrolled, and owners can be notified to secure the property and take measures to prevent fires. This technique was used by the city of New Haven, CN, to help control a serious arson problem during the early 1980s. This was done along with the formation of an arson task force and an aggressive fire prevention program. The subsequent results have been used as a model for targeting arson activity in urban and suburban communities.

The advantages of accurate time information will allow for greater insight to the schedule of the MFS. Accurate time identification will allow the tracker to look at the MFS's availability to set the fires, leading to information on possible work schedule, school recesses, patient or prisoner release, and so on. Combined with location, the time of ignition can help establish travel time and possible method of travel. Time of ignition will reveal clues about the offender's social habits. For example, late night or early morning incidents could identify night dwellers or subjects with sleep disorders. Individuals who are active during these hours are limited in the places they can shop, eat, hang out, and socialize. As stated in chapter 2, time between incidents will lead to clues on motivation. For the interviewer of possible suspects, it will be useful to remember the information about time as an interviewing tool. Simply use the time of the incidents to link suspects who frequent the area during the ignitions. Show the time distance relationship to prove how it would be possible or impossible for a suspect to be in one place or another within the stated timeframe.

We could play the same philosophical rationalization game with days and months. The day of the week and the day of the month or date are the last items required to answer the question "when?" The date of the incident is cut and dry. Without the date this process should not be attempted. The days of the week that the incidents occur on may require some minor effort to gather. Few documentation systems will identify the day of the week. This is not usually difficult to add, but as with anything else, it can be time consuming. Having a calendar handy will make this easier. In chapter 9 we will discuss the further use of calendars in this process. (Before purchasing a calendar, be sure to read chapter 9.) It is also worth the effort to identify where that day falls within the month. This is not to say the date, but which week of that month. For example, the third Thursday of every month may appear to have some significance in the pattern search. Without this identification, the process will only note Thursday as an active day. Not until some further research is done will it become apparent that it is the third Thursday of every month. If the active days are plotted on a calendar, it will definitely be noticed. Unfortunately, unless a calendar is fully utilized, the only thing to note will be that Thursday is the active fire day. The fact that on the third Thursday, brush fires occur around the same time in the same general areas will not be identified.

The analysis of the day and date data will also help to establish the social habits of the MFS. For example, fires during the weekdays and not weekends may help to identify work schedules. Other examples of particular days may identify paydays, using the example mentioned earlier where the third Thursday of every month is popular. This may signify the second paycheck of the month that can be spent frivolously. Another example would be with a certain date or dates that hold some significance to the suspect. Extremist MFSs will act on a specific date in order to give special emphasis to their terrorist acts. Some dates and days may signify abuses suffered or observance of some ritualistic act.

The second item of importance is location. As real estate people say, "location, location, location." Military types say, "choose your battle ground" and preferably "seek the high ground." To answer the question of "where?" we must know the incident location. When gathering information on location, be sure it is accurate. If the incident was identified as TRO (to the rear of) be sure to state the location as such in the data file. The same is true for across from or next to. (Incidentally, the unsub may even be using these same terms or phrases that are common to public safety officials and this may end up being a clue to finding the MFS.) If cross streets do not match, research the alarm further. If no further information is available on the incident, do not use it. When identifying a non-structure fire, try to use landmarks or fixed points. For example, in the case of woodland or undeveloped areas with no visible land marks, use measurements and compass headings, as well as latitude and longitude coordinates from GPS or area maps, for the distance from the nearest cross street. When dealing with MFSs or unfounded calls where the cross streets and the addresses are consistently wrong, check the notification information. This may be another clue to finding the unsub who may be familiar with an area, but does not, as of yet, have an intimate knowledge of the area.

Location will also reveal many clues about the unsub and the future of the fire patterns. Cluster zones or ground zero will usually lead to the suspects and their hang out, hide out, or residence. Knowing the most frequent locations will also assist in the establishment of surveillance.

When MFSs ignite fires in undeveloped areas, they have to get there, start the fire, and get out. During those three phases, there is a potential that a suspect will leave some clues at or near the crime scene. The basic forensic investigative skills will yield the best results in these locations. Within a cluster zone the potential is highest. MFSs are not always consistent on exact time or what they use to ignite a fire until they master their craft, but MFSs are usually consistent on location due to a comfort factor or confidence that has built up over time. In some cases, the MFS may have grown up in this area and spent years learning every bit of the landscape. MFSs also tend to case their target area during their travels, as if they were getting to know the geography on an intimate basis. This geographic knowledge will be put to use when planning to set a fire. They will know the best routes in, the best way out and who is usually around the target area. It will normally not be a haphazard approach. There will be some thought (although usually limited) to these concerns. A good question to ask during the interview process is, "Do you know this area; have you ever been there?" Unless the offender is in an intoxicated state, location will usually have some significance, even if the suspect (or disorganized offender) does not realize it. Even malicious mischief and drunks will tip their hand, to some extent, with location due to the fact that the fire is usually set in the path or direction in which they are traveling. Locations or the distance between location will also help identify the method of travel, or more informally, how the suspect gets around. This goes back to what was discussed earlier in this chapter, the time distance relationship. In order to case a site, it will usually involve some form of transportation. It could be by foot (as when fires are set close together), by bicycle (as when fires are set further apart), or motorized (as when fires are set at great distances from one another). For example, the unsub may rely on public transportation, which is most likely the public bus routes. This may be the unsub's primary form of transport allowing the unsub to gain knowledge on bus schedules and be able to observe target areas from the anonymity of a passenger seat on the bus. A simple example would be a MFA suspect who could use the pay phones along the bus route to phone in incidents while waiting for the bus to arrive. Simply identifying the time between notification locations could be useful. Another application of location in reference to travel occurs when the unsub operates on unpaved roadways. These MFSs will usually follow some type of trail such as power lines, horse trails, and bike trails or fire roads. Therefore, when studying cluster location, be aware of the access and intersecting points. It is rare that a MFS will blaze a new trail in or out of an area. Non-accessible fires are probably not the work of a MFS.

The last item under the question of "where?" would be the community and/or fire district. Some MFSs will concentrate on particular communities and/or fire districts for various reasons. Those reasons will usually relate back to the motive. For example, firefighters who turn to arson will usually ignite fires within their response area. Because action or vanity motivates this type of MFS, they wish to participate in the fire suppression operations. If the fires were not in their response area, there would be no point. Another example of where location can relate back to motive is that of revenge. Students or graduates who feel they were ill-treated by the staff or the butt of abuse by the student body of a school will tend to see the school district as one of the causes for their problems. These suspects may set fire to the homes or vehicles of students and/or staff members. They may also attack school structures such as sheds, out buildings, or dumpsters throughout the school districts' varying locations. Keep in mind that many large schools and regional school districts do not always follow the same community boundaries. These boundaries can be very large, encompassing more than one source from which to gather data. Due to their size secondary schools often contain students from diverse cultural and socio-economic groups. These students form multiple social groups or cliques with varying missions and agendas. The potential for clashing groups or rivalries will almost always exist. This is compounded when the school district severs a large geographic area. In these situations it is not uncommon to find large school districts encompassing multiple jurisdictions. In Suffolk County there are approximately six school districts. One of these school districts covers three townships and includes five communities with five separate fire departments. When analyzing the details about "where?", remember the big picture as well as the specific.

Finally, the last investigative question we need to discuss is that of "what?" More specifically, what type of fire incident? The type of fire activity will be a good clue to the MFS motivation and at what stage or level of violence the unsub is. A graduation in the type of incidents may indicate juvenile or malicious mischief suspects who are working toward elevating the level of violence. They will follow classic stages of graduation, starting with grass, brush, rubbish and/or dumpster fires. These incidents will then escalate to automobiles, abandoned structures, and finally, storage sheds and occupied structures. Some will deviate and may move toward other crimes such as vandalism and theft. But no matter what direction their criminal activity may take, fire will always be part of their résumé. Most seem to find a level of violence to satisfy their need and then move on. This will appear as cases that may have gone cold. Something may change in the life of the MFS that refocuses their attention. In these instances it would be wise to identify what other types of activity has been occurring in and around the fire area. Some MFSs will continue for months or years setting the same type of fire or phoning in consistent MFA calls. It will be with these types of MFSs that the process will work best. For others (usually the sociopath) the ultimate escalation would be to murder. This has and will continue to occur, but only in a small percentage of cases, usually occurring when occupied residences become the primary targets. This is where most traditional MFS cases begin, usually under the heading of serial arsonist. Fortunately, the majority will not escalate to this level.

The types of fire(s) set will usually tend to be the first clue to something unusual. This is especially helpful if the individuals who are conducting the track are unfamiliar with the locations. For instance, when the HI report states something along these lines "I cannot tell exactly where they happened, but I can tell you there have been a large number of car fires in that community." In this instance, the similar type and locations of the fires can be noted so as to identify the motive of the MFS. Another example is fires occurring at government facilities may link incidents to anarchists or extremists. As covered in chapter 2, these groups have specific goals. Their targets are almost always pre-arranged and planned out for maximum effect. They will look for something that symbolizes what it is they are against. This does not always mean an occupied structure. It could be sheds or storage buildings used to house communication equipment, supplies for research, or record storage. Because most of these offenders will choose targets distant from one another, some time will need to be spent researching past incidents outside the normally considered cluster zone. Although beyond the standard perceived geographic distances applied, these incidents will contain similarities. One such similarity may be the use of a device or method of entry. Also, the method of ignition may or may not change with these types of incidents. On a positive note is the fact that, in some cases, the suspects will announce themselves or take responsibility for their actions immediately following the incident. However, recently many of these so-called leaderless or decentralized groups have been taking less and less responsibility for their actions. As covered in chapter 3, be sure that the type of incident is specified. Do not confuse one type of fire with another type of fire.

These are only a few examples of how types of incidents can be related to possible motives and suspects. There are many more. The tracking process is not designed to answer all the questions surrounding the suspect's motives. Instead, it is designed to bring the tools together to help identify the track and allow others to further research the results.

Chapter 6

HOW TO ANALYZE THE INFORMATION COLLECTED

To begin the discussion on how to analyze information, we will identify the perfect set of circumstances such as what you might find in a perfect world. First, you would need access to a crime analysis unit. This unit would be fully staffed and open to conduct new searches. By simply explaining what would be required to staff members, the analysis unit could develop the database, find the data sources and acquire the expert help needed to process and analyze all the information. This crime analysis unit would employ state of the art computers with the sources of the data required already stored within the system or networked to the mainframe server. If this unit were not available, then the data alternative sources would be willing to supply a diskette of the data in a program that is compatible with the database used by the crime analysis unit. By accomplishing this, the investigator or tracker would be able to save many steps and a great deal of the time required to enter, sort, collate, and produce the search groups needed for tracking. Note that the term "if" plays a big role in a perfect world scenario. In my experience of conducting the tracking process, there has never been a perfect set of circumstances.

In regard to the thought of a crime analysis unit being available to conduct this type of search, this is more of a dream than a reality. Even if a unit such as this existed, getting the end results, a completed FITR (fire incident tracking report), in a timely fashion, is unrealistic at best.

This is not to say that these units do not exist or that you will not be successful with the ideas presented in this perfect world scenario. For example, investigators with the Overland Park Police Department in Kansas already have integrated GIS software, as well as a crime analysis unit with a full-time staff trained and equipped with the technology to be fully operational. Hopefully, as the future brings computer technology to more disciplines, the administrators will realize the value of these units. In turn, the development of these units will create the opportunities where this tracking process would become a common procedure.

If you are successful in your endeavor to obtain such tools as GIS, be sure to review all sources of information so as to eliminate working with improper or unnecessary information. My limited experience with GIS systems has produced less than desirable results (this will be covered in more detail in chapter 7). For now, realize that unless the data is properly formatted or geo-coded to the GIS system, it will not always be available for analysis. I have found that, at this time, the data received from your sources will usually not be in a format that is compatible with your program. I have tried to transfer data files on one or two occasions. The result was picking up the nickname "Virus" (short for computer virus) from my coworkers for messing up the office computers every time I have attempted it. Consequently, as for the supply of source data, most of the time I have been happy to receive a printed copy to avoid sitting in front of a computer screen transcribing all the data to paper by hand. In instances when the data is received in the same program format being applied, the data will likely contain items that are not needed for analysis and will only slow down the process. In these instances the additional information should be excluded from the entries. The reasons for doing this will also be discussed further in chapter 7. For now, please consider the recommendation of excluding them from the entries.

Your original source documents should be retained with the additional information. The simple function of data entry will not only help remove unnecessary information and catch errors, it will also help to ingrain certain locations, types, and times to memory. This will be of assistance later on when someone else in your unit, agency, or other disciplines begins to discuss an incident with similarities in locations or times to the ones being analyzed. Remember it will be rare that any one individual in an agency will have knowledge of all the incidents. Data entry will also help if you get ambitious enough to listen to the fire and/or police frequencies during the hours and days of activity. Remember that you are one of the best assets available as an intelligence source.

Once the data entry is completed, the next step will be the need to query the information into usable search groups for analysis. It is during this analysis correlation that the work begins to get interesting. In some cases, patterns will begin to jump out at you. This is the point at which it is important not to get ahead of yourself and start reading into the data. The next chapter will discuss in detail how to avoid errors. Remember, let the facts speak for themselves. As Detective "Joe Friday" would say, "Just the facts, Ma'am." Do not jump the gun or go off half-cocked because something appears to be happening. Remain objective and follow through to be sure of all the possibilities.

To begin the query process, the information should first be broken down into some search parameters. These parameters should be built on the individual items identified in the collected source data. These individual items will make up the fields of the computer database file. Whether the data is sorted by hand or by computer, the breaking down of the information into fields will be the key to successfully conducting the search process. You should label these fields to coincide with the type of information stored in each field. The recommended titles for the fields are: time, date, type, location, fire department, area (meaning the community), week of the month (simply referred to as "WEEK"), and any additional special notes (referred to as "NOTES"). The fields for time and day are self-explanatory. These two fields can be titled "TIME" for the incident's time of occurrence as discussed in chapter 5 and "DAY" for that specific day of the week.

For the type of incident field, it is recommended to use the title of "TYPE". This will refer to the specific type of fire that occurred; for example, automobile fires could be abbreviated as auto and structure fires to "struct". As with all the fields, be sure to place the proper details in the proper field.

The location field can be titled "LOCAL". This field should contain the address, latitude, and longitude or the best reference points available to the exact incident occurrence (this will also be covered in more detail in chapter 7). Whether by hand or computer, try to arrange the location information so that it can be easily correlated. For instance, in a street address it may help if the street name precedes the alphanumeric symbol. Placing the alphanumeric symbol last will allow the analyst to quickly scan the list for the same street name (for example, instead of 1710 South Olive, enter South Olive, 1710). This same idea may be useful in the computer version for similar reasons. With most computer database software, the data can be sorted in ascending or descending order. Most database software will use the first symbol in the designated field to assign the sorted position on the list. If the first symbol is the numeric address, then the sorting order will be by number and not by street name. Hence, the advantage in placing the street name first.

The next field to discuss is the special notes. The recommended title for this field is "NOTES" or "NOTE". Special notes would include names, cross streets, and evidence. Names would include that of the caller, the resident of the structure, subjects noted in the area, and witnesses. Cross streets are important especially when discussing structures, dumpsters, and car fires. For instance, while I was investigating a rash of car and dumpster fires in a community unfamiliar to me, the first thing that jumped out when entering the data was the cross street. All the incidents occurred along a two-mile stretch of a north/south main road. This main road was the cross street of at least 60% of the incidents. Once the incidents were plotted on a map, it was noted that they all fell within less then a half mile east and west of the main road. (If the names of the cross streets are available, a separate field can be created for cross street information as well. However, I have found it preferable to place the cross street information in the "NOTE" column). In most cases simply place the cross street information on the end of the street address with the special items going in the "NOTE" field. Evidence at scenes that consistently shows up should also be placed into the "NOTE" field.

A word of caution is appropriate here. Although recurring evidence should always be noted, do not get bogged down with great details about the evidence. Try to abbreviate, writing only that something significant was noted and that further details are available in the fire report. As mentioned earlier in this chapter, items can be overlooked or missed at future scenes if not noted in the data. Do not get so focused on the cause and origin evidence that you tend to overlook other trace evidence. This other evidence may also have been noted and collected at other crime scenes.

When analyzing data within the jurisdiction of a specific fire department, it will not be necessary to specify the name of the fire department for each entry. One entry for that field can be used to identify the data. The next field to discuss is the area field, which refers to the community where the incident occurred. A recommended title for this field is "AREA". If the fire and/or police department districts searched have the same name as the community, you can just use the term "FD" or "Dept". On the other hand, when there are incidents in more then one district or community, the specific name of the community, fire department, or police department should be included for this data.

Depending on the record keeping and/or incident reporting system within the jurisdiction in question, it may be necessary to add a field for case numbers. As mentioned in earlier chapters, some of these incidents may not have been investigated due to the nature of the incidents. As the pattern begins to form and the track develops, incidents may need to be assigned case numbers for further investigation. A case number field will be helpful for keeping track of the incidents in the future as the record log will no longer be in sequential order. The case number field may also be helpful if the track involves multiple jurisdictions with similar cases developing within the individual agencies involved. As additional information is developed, it will be helpful when relating that information to a specific case and a specific investigator in a specific agency. The case number field can be simple titled "CASE". Do not confuse this number with the number assigned by the database program, sometimes referred to as the record log. Most database programs will automatically assign numbers in sequential order of entry for incidents added. This database number can help identify the total number of incidents studied but should not be used for incident numbers, as it will be unique to the program and not the overall investigation.

Now that we have covered the types and title for each field and what data will be entered into them, we can discuss the search analysis process. Each search will consist of all the fields with at least two acting as the key fields. These key fields will make up the first two items in the search. These key fields will also be the titles for each search group or query in computer language. The search groups will be analyzed for information on the unknown subject's patterns. Each search group arrangement will vary so they will be covered individually. It should be pointed out that this is not the only way to conduct the search process. Rather, this is only the recommended way to conduct a search analysis. The reader is encouraged to try any method of search that reveals the best results. The only thing that remains constant in this process is the data required.

Having a computer will make the process much easier. A GIS program with compatible database software helps reduce the process time dramatically in most cases. If there are no computers available, the process will be much slower. This is not to say that the job cannot be accomplished. However, it will require you to spend more time in data entry, collating, and sorting with less time to do the analysis. For those who are not yet using a computer, this chapter is arranged so that the beginning of each section discusses the manual or handwritten method involved in the process and then moves on to the steps involved in the computer method. It may be helpful for those who are using a computer to also read information on the manual method to understand the process better. Decide for yourself.

Whether you have a computer or not, first collate your data. The recommended way to start is by looking at the date and the time of the incidents as your first key fields for the search groups. Sorting is one of those time-consuming parts to this process. Depending on the number of data sources that are involved, this could take an hour or two to complete. Once arranged and entered, you can move on. When sorting by hand, be sure to make additional copies of this search group list. Those copies will help later when developing other search groups.

The first search group discussed is referred to as the "DATE and TIME" search. By searching the date and time groups first, it should be possible to confirm much of your HI sources that you may have. If there is any activity that appears to be consistent or excessive, it should be noted. The DATE and TIME search will also help with the narrowing of the searches to possible months and days that the unsub (unknown subject) prefers. Specific months may give some leads to the MFS activities that relate to work schedule and/or employment status. This will help you identify an organized or disorganized offender. The DATE and TIME search will be the easiest to update; simply add the most recent incident to the bottom of the list. The first maps generated will probably be based on this search. The DATE and TIME search group will also be used for calendar plotting. It will help in the decision to use a one-year, six-month, or a quarterly month chart. This search group also helps identify the dates that the cluster center and ground zero events had occurred. The DATE and TIME search group file is arranged in the following order of key fields: Date, Time, Day, Week, Type, Location, Notes, and Area (see Table 6-1). This table displays the basic design for all search groups whether the search is done manually or by computer.

Table 6–1:

DATE	TIME	DAY	WEEK	TYPE	LOCAL	NOTES	AREA
12/3/03	03:11	Sat	1st	Auto	5th Ave. and Main St.	Abandon	Littleville

DATE and TIME Search Group

The second recommended search group is "DAY and TIME". When conducting the search by hand, use the additional DATE and TIME search group copies to assist you. It is recommended that more than one copy be used to eliminate the risk of color coding confusion. Although it may not seem to be a problem now, it will be if you try to assign more than one function to a particular color. First, have a colored pen or highlighter handy. Second, go to the "DAY" column and then go down the column sorting all individual days. Start with "Sun" or Sundays and mark all Sundays with one color. Once this is completed, the third step is to place the sorted days into chronological order. Go to the "TIME" column next to the "DAY" column and look for the earliest Sunday incident. Sort all of the incidents so that the new list will start with Sundays as the first item in the "DAY" column. The very first Sunday will be the earliest incident in chronological order. Once transcribed, cross the incident out or place a check next to the time, whichever you prefer. Continue this function with all seven days of the week. If necessary, make additional copies of the "DATE and TIME" search group for every two days completed. Upon completion, the DAY and TIME search group will be established. Querying the data from the "DAY" and "TIME" fields will do this computer search. The DAY and TIME search will be most helpful when developing the final report, which will be covered in chapter 10. This search will help establish the following: busiest day, busiest time of that day, which days see multiple incidents (if any), which days are least likely to see activity, and where most activity occurs on a particular day. This search group may also help toward identifying the suspect as an organized or disorganized offender when a time-frame correlation is noted with the incidents. For example, does the unsub prefer specific times on a Sunday or is the unsub active Friday evening and Saturday mornings? This could show that the unsub has no weekend commitment and may prefer to tie one on before setting fires. The unsub may work nine to five on weekdays with weekends off. Fire activity through the week during early morning hours may show an unemployed or unskilled laborer (someone who does not require much sleep or without a regular start time for work). Fires that occur through all hours of the day or during specific seasons, may be the work of juveniles. This search group may also help identify the development of cluster zones or the further development of skills and target selection in the case of the organized offender. Variants of this information will include noted weekend activity or specific weeks that the majority of incidents occurred. If the data is accurate, this search can reveal the majority of information from which patrols and/or surveillance will be established. The recommended layout of this search group file is in the following order: Day, Time, Type, Location, Notes, Date, Area, and Week (See table 6-2).

DAY	TIME	TYPE	LOCAL	NOTES	DATE	AREA	WEEK
Sat	11:03	Auto	5th Ave. and Main St.	Abandon	12/3/03	Littleville	1st
Mon	23:15	Dumpster	325 4th Ave.	x/St. of 5th Ave.	12/5/03	Littleville	2nd

Table 6-2: **DAY and TIME Search Group**

The next recommended group search is the "TYPE and DAY" and if possible "TIME". To develop this search by hand, once again use copies of the DATE and TIME or the "DAY and TIME" search groups. Using the same technique as in the DAY and TIME search, have a highlighter ready. First, pick the type of incident you would like to start with, possibly the most frequent, and highlight those incidents. Second, follow the same technique used in the DAY and TIME search; the only thing different with this search will be the need for additional copies of the DATE and TIME search group. Once the search list is completed, make copies for use later. As for the computer query, simply set up a data run with the type, day, and time as the primary key fields. This group will show the most popular type of incidents on particular days (if any). It will also show if there is any correlation with particular incidents on particular days. If time is added as a key field, this search will show the most frequent type, day, and time that these incidents occur. For example, some subjects choose to set dumpster fires on weekdays and structures fires on weekends. Perhaps, it is further noted that these dumpster fire, not only occur during the afternoon hours on weekdays, but also during early evening hours and the structure fires occur on weekends during evening and early morning hours. This analysis can help us to focus on juveniles or unsubs who must be home by a certain hour during the weekdays. Hence, from the TYPE and DAY search it may be possible to identify when particular types of fires occur on specific days in particular locations. When placed in the recommended order, it is possible to search the location, as well as to identify the type, day, and time the incidents are occurring. This search can be used to cross-reference the DAY and TIME search to confirm or disprove findings. Because the majority of MFSs follow the standard escalation of fire setting or remain consistent, we can analyze this search group to see if the activity is consistent with what has already been observed or if it is escalating. However, although the day and time of the incident may match other incidents, the type and location fields show no similarities. Unless the fire investigator's evidence directly links a particular incident to the others, these incidents should be considered as possible, but not probable.

The TYPE and DAY search also helps to create a map to show where particular types of fires are occurring, as well as cluster zones by type of fire and days. Unless the personnel conducting the analysis have a tremendous knowledge of the active area, map plotting will be necessary. (Mapping will be covered in chapter 8. This is one of the final actions to confirm the presence of the MFS activity.) Once the presence of a MFS is confirmed, the type and day information is used to establish the type of patrols and surveillance needed. For example, authorities could use unmarked vehicles near a residential development where rubbish fires occur at specific times or use marked vehicles at night in unpopulated areas where car fires occur. The TYPE and DAY search group can also be used to help in the development of a fire activity report that can be used by fire suppression units to assign responses. This is especially helpful when numerous calls occur on the same day. For example, if fires are occurring in groups of three, with 30 to 60 minutes between each, and the third fire is most likely the largest (possibly a structure), then assets can be held in reserve before being committed to the first two incidents (which are usually minor). The recommended layout of this search group file is in the following order: Type, Day, Time, Location, Notes, Date, Area, and Week (see table 6-3).

TYPE	DAY	TIME	LOCAL	NOTES	DATE	AREA	WEEK
Auto	Sat	11:03	5th Ave. and Main St.	Abandon	12/3/03	Littleville	1st
Dumpster	Mon	23:15	325 4th Ave.	x/St. of 5th Ave.	12/5/03	Littleville	2nd

Table 6-3:
TYPE and DAY Search Group

Another useful search group is the "TYPE and DATE" (I personally think this search group requires the least amount of analysis). When arranging the parameters of this search group manually, use the table set up for the "TYPE and DAY" search group. (The original "DATE and TIME" search group can also be used, but will require colored pens and/or high-liters.) The use of the TYPE and DAY search group appears to require the least amount of analysis. First, take the TYPE and DAY search group list and draw a line at the end of each incident type established. This will help to distinguish between each type of incident (see table 6-4).

TYPE	DAY	TIME	LOCAL	NOTES	DATE	AREA	WEEK
Auto	Sat	18:33	15 Main St.	x/park	9/3/03	Littleville	3rd
Auto	Thur	00:33	5th Ave. and Main St	abandon	9/30/03	Littleville	1st
Auto	Wed	21:00	Grover La.	x/st. Main St.	4/13/03	Littleville	4th
Dump	Fri	23:00	3 Lake St.	x/st. 8th Ave.	6/5/03	Littleville	1st
Dump	Sat	01:30	8th Ave.	TRO Drugstore	3/11/03	Littleville	2nd

Table 6–4:
TYPE and DAY Search Group with line between types of incidents

Next, go to the "DATE" field. Transcribe the incidents (by type) to a new list in sequential order starting with the earliest date. Be sure to leave space for additional entries under each type. As with the DATE and TIME search group, this list can be updated by simply adding the most recent incidents to the bottom of each incident type. This search group should be arranged in the following order: Type, Date, Time, Location, Notes, Day, Area, and Week (see Table 6-5).

TYPE	DAY	TIME	LOCAL	NOTES	DATE	AREA	WEEK
Auto	Sat	18:33	15 Main St.	x/park	9/3/03	Littleville	3rd
Auto	Thur	00:33	5th Ave. and Main St	abandon	9/30/03	Littleville	1st
Auto	Wed	21:00	Grover La.	x/st. Main St.	4/13/03	Littleville	4th
Dump	Fri	23:00	3 Lake St.	x/st. 8th Ave.	6/5/03	Littleville	1st
Dump	Sat	01:30	8th Ave.	TRO Drugstore	3/11/03	Littleville	2nd

Table 6–5:
Handwritten version of TYPE and DATE Search Group (Note the space between types that allows you to record additional incidents)

The computer query will only require that you properly arrange the fields so that TYPE is first and DATE is the second field. This search group will help identify what level of violence an MFS is operating. The TYPE and DATE search group will show fire setters who may be satisfied with continuing to set rubbish, brush, and/or nuisance fires for their entire career. For example, a 56-year-old subject was arrested in Lubbock, Texas, for setting approximately 140 fence and juniper bush fires in a community. The offender was an accountant who seemed to enjoy seeing things burn. His fires very rarely threatened any structures or individuals. Another 56-year-old subject would set garbage cans on fire when he was unsuccessful in seducing a sex partner. He would only set the one can on fire after the failed seduction. The subject would then be aroused by the flames and masturbate. Both subjects had conducted this fire setting activity for well over a year before being noted. As for the other MFSs, this search group will show the progression of the more violent offenders. For the more violent MFS, the fire activity will escalate as if it were following a scripted order. Due to this trait, the TYPE and DATE search group can also be used to weed out non-probable incidents. For example, it is not likely to see two occupied residential fires followed by multiple brush fires followed by multiple dumpster fires unless the unsub sets fires based on the season, which this search group will also note. Some MFSs will ignite grassland fires in late fall or early spring knowing the vegetation is driest and can do the most damage. For future reference, David Berkowitz wrote specific information about the fires that he set. If this is the situation, take note to see if the unsub also uses the geography and wind direction to his advantage to create the most damage. John Orr started his fire service (and arsonist) career as a vegetation inspection officer. His job was to identify areas where vegetation was ripe for burning so that a controlled burn could be done to remove the hazard. The only problem was that most of the identified locations burned before the controlled burn could be conducted. Based on this incident, it would be good to look beyond the immediate community, based on the fact that an organized-type offender may research his targets thoroughly. Fire suppression personnel can use this information to prepare their fire staffing for future seasonal activity. Suppression personnel should incorporate weather conditions during the active fire period. Remember, if these offenders have gone to great efforts to accomplish their goals, all aspects should be considered.

As stated earlier, other searches can be conducted with varying success. The primary items to extract are the possible establishment of patterns involving specific days, areas of fire activity, times that activity may be occurring on specific days and/or in certain communities. An example of other search groups would be based on type and location. Generally speaking, I have not conducted any further research on this issue because the other search groups mentioned have provided the information necessary to complete the pattern search. However, fire prevention officers could use this particular search group to locate and inform members of other disciplines about probable target properties. These properties would be susceptible to fire and would be considered target hazards in and around the cluster zones or active fire areas. Fire prevention officers could also use this search to notify property owners when situations other then MFS have been identified. For example, multiple AFA incidents in an area could be identified and addressed. These alarms may be due to water pressure surges in the area, electrical problems or telecommunication failures. Keep in mind that no matter what the design, these search parameters will be used to develop the final report that is distributed to the troops.

During some searches, it may appear as though you are not seeing the full picture and that more activity should have been noted. If you think that some incidents related to the case may date back further than the original search or that there may be incidents occurring outside of the community that you are analyzing, then further research may be necessary. In situations where there is a possibility of activity prior to the designated search date, these additional incidents will usually be considered unconfirmed or unverified. When this is a possibility, try to identify and confirm any additional information that may be pertinent to your case.

Methods to accomplish this task would include identifying fire activity by the types of fires that are occurring in specific geographic areas on specific days and at specific times. In this way you can also identify unverified incidents, such as uninvestigated dumpsters, brush, grass, rubbish, cars, and structure fires. Simply use a method based on the patterns identified in the search groups. Start by taking the similar traits identified in the patterns and matching those traits against the unverified incidents. Use a minimum of three similar traits from the eight fields in each search group of the confirmed incidents. Using similarities from less than three fields will leave the incident search suspect to question. If there are more than three similar traits, it is better to include all of them in the search. However, do not try to compare all the fields of the confirmed incidents to the unconfirmed, as this will tend to limit or restrict the search from a wide variety of incidents. If necessary, develop a separate search group for the unverified incidents based on the similar traits of the verified incidents.

For example, suppose similar certain traits are noted in the confirmed suspicious fire incidents. One identifies 11:00p.m. to 12:30a.m. timeframe as the most probable time for car fires to occur near 5th avenue (see table 6-6).

TYPE	DAY	TIME	LOCAL	NOTES	DATE	AREA	WEEK
Auto	Sat	00:03	5th Ave. and Main St.	Abandoned	12/3/03	Littleville	1st
Auto	Sat	23:03	5th Ave. and Maple St.	TRO	8/3/03	Littleville	1st
Auto	Sat	00:35	4th St. and Main St.	Abandoned	11/17/03	Littleville	2nd
Dump	Mon	23:15	325 4th St.	x/st. of 5th Ave.	12/5/03	Littleville	2nd

Table 6–6: **Verified Incidents**

You can now set up a search group for the unverified incidents based on the key fields of "TYPE", "TIME", and "LOCATION". Place special emphasis on the "NOTES" field for additional cross street information (see table 6-7).

TYPE	TIME	LOCAL	NOTES	DAY	DATE	AREA	WEEK
Auto	00:33	5th Ave. and Main St.	Abandoned	Sat	9/3/03	Littleville	1st
Auto	18:33	15 Main St.	x/park	Thur	4/13/03	Littleville	3rd
Auto	23:36	Grover La.	x/St. Main St.	Fri	9/30/03	Littleville	4th
Dump	23:00	3 Lake St.	x/St. of 8th Ave.	Fri	2/5/03	Littleville	1st
Dump	23:15	325 4th St.	x/St. of 5th Ave.	Mon	4/5/03	Littleville	2nd
Dump	00:04	880 South St.	x/St. of 5th Ave.	Sat	11/12/03	Littleville	3rd
Dump	22:44	1000 South St.	TRO x/st. of 5th Ave.	Sat	10/22/03	Littleville	3rd

Table 6–7: **Unverified Incidents**

This new search group will allow for the simple comparison of similar traits among the verified and unverified incidents. This particular search group would identify that dumpster fires and rubbish fires have also been occurring to the rear of the locations noted around the same timeframe on Fridays and Saturdays. These unverified incidents can now be considered probable targets of the MFS. Based on this additional information, it could also be concluded that Friday and Saturday nights are the most likely days for fire activity to occur.

If the search was narrowed to only look for cars on Saturdays between 11:30p.m. and 12:00a.m. near 5th Ave., it would have missed the dumpster activity on Fridays. As you can see this method requires the right blend of fields–not too few and not too many. It may therefore require some trial and error.

When fire activity appears to show signs of occurrence outside of the original geographic search area, certain additional techniques can be employed to help discern the level of probability that the same unsub is setting fires in this other area also. In the case of confirmed or verified target communities where these patterns are being noted, conclusions can be drawn about each search group from which comparisons of the patterns can be made with other communities. These comparisons would be primarily based on the "TIME", "DAY", and "TYPE" fields. Over long periods of time, even years, dates can be calculated into the search. Whenever possible, the unverified incidents can be further confirmed by taking a look at the total number of fire incidents reported for that community during a particular timeframe. The timeframe in question could be a year, six months, or three months. Once the timeframe is established, compare it to the same timeframe for the prior year's activity period. Try to avoid using a timeframe of less than two months. The reason for this is that the comparison will be open to many variables. If possible, compare the fire activity during a specific season.

If using the method explained to search a previous timeframe would be too difficult, then a search might also be conducted by using a similar geographic area for the same time period. To help accomplish this task, use the computer mapping software that is discussed in chapter 8 or a plotting map for the hand-produced version. Rather than attempting to re-enter all the data into new search groups, simply take the data for that time period and plot them on maps. Once plotted, compare the activity for similarities. If a GIS system is being used to conduct the analysis, a search of that specific geographic area can be done quickly. When no GIS system is being utilized, the search method will be difficult to apply and will be time consuming if not done by computer mapping. Having said that, if there is a MFS at work, the number of incidents should show a variation in activity. If the frequency of activity is similar in both timeframes, it will either require research of a similar time period further back or the assumption that this is the normal fire activity for the area during that time period. When the latter appears to be the case, the information should be shared with fire suppression personnel, who may not be aware of this pattern. This will allow them to keep a continued vigilance on the situation. Preferably both methods should be applied, but if it is only possible to apply one, then the unverified incidents could be considered for the list as probable activity of the MFS. When at least two methods are used, the confidence level would be considered high enough that the unverified incidents should be added to the verified incidents of the MFS.

The pattern developed from the search parameters is based on the facts surrounding the incidents and can then be used to prove or disprove information gathered from those HI (human intelligence) sources. Once this is accomplished, all the valid intelligence will be used to establish a track. This track and any other valid intelligence will be used to establish a fire activity report or FITS (fire incident tracking system) report. When subjects are identified it can be used as probable cause for investigators. The final report will state whether a pattern of fire activity has been identified or that. Based on this track, conclusions will be drawn and written into the report that will be distributed. As future incidents occur, the data will be easy to update. The patterns can then be reviewed and tracking will be improved as the reports are refined.

Chapter 7

How to Avoid False Tracks

One of the worst ways to provide fuel to the doubters and skeptics of this process occurs if the information distributed is inaccurate and not useful. To prevent this from happening, it is recommended that you avoid using bad information. Bad information input will lead to bad information output. For this reason and a few others, this entire chapter is dedicated to preventing bad analysis, which creates inaccurate patterns that lead to false tracks. Bad information comes in many forms–bad tips, bad addresses, and incorrect records are only a few examples. Although these are some of the most obvious problems, they are not always the culprits of bad patterns and false tracks. We will try to look at most, if not all, the reasons for false tracks. The study of these problems will assist with the development of a checklist to identify items when entering and analyzing the data for accuracy. Make no mistake–this will take additional time, but it will be necessary, especially to ensure the validity of the track.

When dealing with a large amount of information, the simple act of proofreading the completed work will present the greatest challenge. Sometimes, too much information is not always the best way to work. One of the biggest problems with the United States intelligence gathering services is not that they receive all bad information, but that there is too much information gathered. In fact so much information is collected that there are not enough hours in a day to analyze it all. This has led to many missed opportunities and incorrect analysis. The same can occur with fire incident analysis if the proper parameters are not established. It would be great to know all the specifics about the incidents being analyzed, but it will cause the person(s) conducting the search process to bog down in information overload and analysis paralysis. This is where some GIS systems can confuse the process. With other crime analysis processes, such as robbery or rape investigations, GIS can be a great time saver when data is input properly.

Unfortunately, fire incidents do not present the same luxury. First of all, the incidents must be determined to be suspicious (which not all are). Secondly, not all are investigated even though they may be suspicious. To make matters more confusing, not all fire incidents are investigated by one agency in the same community, thus causing development of varying data. Whereas an environmental or wildlife law enforcement agency may investigate wildland fires, local fire and police departments may investigate car fires, and regional, county, or state law enforcement agencies may investigate fires involving fatalities. Furthermore, incidents involving houses of worship may fall under federal law enforcement jurisdiction. Yet, only one suspect—the MFS—may have set all of the fire incidents mentioned. Most GIS systems will not have the flexibility to interact or quickly take data from all of these agencies. More importantly, all of the data that these agencies collect will not be required. Do not be compelled to gather more information than what is needed, even if the software can extract data quickly. The National Fire Academy's AIMS (Arson Information Management System) or the FBI's VICAP software should be used for detailed record keeping. The AIMS and similar software are much better suited to enter and retrieve the details about specific incidents. VICAP software has been specially developed to gather and collate data on violent crimes. For the purpose of this process, we are looking for quality, not quantity. As stated before, keep the information brief;

use the "NOTES" field for additional information, but limit it. For example, when noting fire activity at an abandoned structure where a flare was believed to have been used as the ignition source, write "Abandoned, Flare found." Do not add information about the type of structure because that is noted in the "TYPE" field. There is no need to go into detail about where the flare was found or the type of flare. If it is likely that others will be reviewing the search groups or raw data, simply write in the "NOTES" field "see file" or note the incident report number in the field so that they have some reference point. This will notify other individuals where to look for the additional information. In situations where the analyst is the only person reviewing the data, this will not always be necessary, unless the incidents are so numerous that they are difficult to collate or they number into the hundreds or span years of activity.

For many serial offenders, their first acts of violence involved fire. By tracking these fire incidents, the location of future offenses may well be identified. However, the tracks of an MFS may span many years and can be mixed in among hundreds or thousands of other incidents. Based on the number of incidents to review alone, sorting through these incidents would take many man-hours with information overload occurring. How can the process be sped up and information overload avoided? Simply use the method discussed in chapter 6 and displayed in Tables 6-6 and 6-7. Before back searching those enormous incident files, first look at the known suspicious cases from both the current and over the previous years. Do not look at cases that are undetermined or non-investigated incidents. This approach will hopefully keep the suspicious incident list short. Next, take the suspicious incidents and search for consistent traits between those on the list. Use the search groups covered in chapter 6 to look for similar traits such as consistent days of the week, specific periods of time on those days, or specific dates of the month. Once these traits have been identified, apply them to the current list of incidents within the original search timeframe. Once again, remember to try to find at least three separate traits that show some consistencies. If more than three can be matched, all the better. However, if less than three match, it would not be a good idea to add the incidents to the pattern analysis. One of the least effective consistencies is location. Remember, fires do happen; just because a fire occurs in the believed cluster zone and falls within the timeframe does not make it an a MFS crime. Try to match as many of the fields in the search group as possible, but do not try to connect the dots to something that is not there.

Although instinct is probably one of the best skills an investigator can have, especially when interviewing, do not try using it when tracking. When processing the data, stick to the facts; only consider an incident based on the facts. If you get a gut feeling about an address or type of fire, note it, but do not add the incident to the data list until it can be cross referenced and verified with other resources and/or data bases. Do not read into the information anything that is not there. When a tip is received from a source, do not enter it into the database without specific details. Take note of it and then verify it. Just because someone says something was suspicious, do not count it until it is verified with the primary source or the official incident report. Whenever possible, consult investigators who are handling the case. For incidents where there was no investigation, check with the fire department that extinguished the fire. Remember, in many cases MFS activity may only be noted after the subject has graduated to setting structures; thereby, you are missing all the prior fire activity, such as rubbish, brush, and car fires. This also goes for cluster zones. Using a map of the affected area, draw a circle around the active fire area. Then check to see how much fire activity was noted within the encircled area prior to the original search timeframe. The size of the encircled area will depend on the geography, demographics, and the method of travel used by the MFS. This subject will be discussed further in chapter 8.

As mentioned earlier, when deciding how far back to research, try to start back at least two months. When a tip is received identifying a particular area, check the fire activity for the last six months to a year. During the researching of the incidents, if it is noted that the activity preceded the original search, it is recommended that an additional two months or more be checked. Once the search period approaches a year or more, it may be necessary to alter the search methods. Although it may have not been originally clear during the primary search because the incidents in question may have been part of normal fire activity. To confirm these possibilities, compare the similarities in the timeframe or duration of fire activity. For example, if the original search started with July and went back to the previous January, compare the duration with the six months prior to January of the previous year. For example, compare the time period of January 1998 to July 1998 with the time period of August 1997 to January 1998. If it can be accessed, look at the same period or duration for the prior year. Once again, GIS software can usually expedite this method of search, but do not be afraid to try searching by hand.

The reason to establish if the unsub has been operating for a number of years is to identify the original hunting ground or ground zero and possibly tie the unsub to his or her home area. People have detailed knowledge of certain areas and little knowledge of other areas. With this knowledge (or lack thereof) comes a degree of confidence in a person's ability to move in and about that area with few problems. Areas become more familiar as they are frequented more (*i.e.*, the route from home to work). For criminals, the areas with the highest level of comfort turn into areas in which to commit offenses. And although there are exceptions, most serial offenders have a starting point (*e.g.*, home or work) from which the attacks are initiated. For instance, according to Michael J. Cabral, Deputy District Attorney for Los Angeles County, when the MFS, John Orr, was finally arrested, the total number of fires in the Glendale, CA, area where he operated dropped by 75%. He is believed to have set over 2000 fires during his career with the Glendale Fire Department. Los Angeles County was his hunting ground.

Keep in mind that once the MFS or any serial offender becomes confident, he or she will venture beyond their original base of operation and continue to expand their hunting ground. Attention (via the media or community awareness) causes the unsub to retreat back to the original base hunting ground. This is the area where an unsub is most comfortable. This fact was discussed in the CMRC (Crime Mapping Research Center) paper entitled "Geographic Profiling". It was there noted that "an important geographic property of serial offending is the rarity that an offender will attack close to his/her starting point. This means that the lack of an offense in an area is as much a spatial clue about an offender's starting point as the crime site locations. Another geographic property is that on average, serial offenders will attack within a fixed distance from their starting location."

In situations where numerous fires have occurred, but have gone uninvestigated (such as when rubbish, dumpster, and brush fires all occur in the same area), check to see if the same amount of activity has been noted in the surrounding communities or communities with the same demographics during the same timeframe. For example, while conducting a search, a cluster was noted involving car and structure fires surrounding a railroad station over a one-year period. When I mentioned this fact to a fellow investigator in my office, he responded by saying that this would not seem uncommon for that type of community. To be sure, I researched six other communities with railroad stations that were geographically similar to the one in question. Two of the communities were more affluent and less populated than the community in question. Two more communities were of the same or similar affluence and population as the community in question. The last two communities were less affluent and had larger populations than the community in question. All were checked for the same time period, which meant they experienced the same weather conditions, current events, and difficulties in lines of communication, that is, all automotive road conditions, rail traffic, aircraft transport conditions, pipeline product transfer, hard-line, and wireless telephone reception. The reason for noting this is based on the fact that if any natural or manmade disaster occurred it would not be isolated to one community and would effect most, if not all, the communities to some extent. None of the six communities searched showed the same or similar fire activity to the one in question. This did not prove that all the fires were the work of an MFS, but it did show that something was unusual or unique to that community, which justified further investigation. As the MFS John Orr stated during his trial, "There is no such thing as a coincidence in an arson case."

As mentioned in the beginning of this chapter, bad addresses and incorrect record keeping are two of the first items that come to mind when data seems inaccurate. Bad addresses or records should be caught when the data is cross-referenced with another information source such as one of those discussed in chapter 4. The importance of cross-referencing needs to be stressed here, especially when relying on GIS software exclusively. This is one of the few cases where the manual method will have the advantage over the GIS processing method. The reason for this is based on the fact that many GIS programs will not recognize certain data unless it is entered or input in a specific fashion. When more than one person is entering the data, the potential for variation or different techniques of data entry increases. Something as simple as using all capital letters or forgetting to spell out the type of roadway can cause havoc.

Additionally, some GIS programs will recognize abbreviations, but not all. For example, placing TRO (to the rear of) in front of the address would cause the data to be misinterpreted and dismissed. Consequently, when the GIS software attempts to place the incidents on a map, many of the incidents may not be processed. We noted this problem with our GIS program when on an average 45 to 50 percent of the data was not being plotted. We had to go through each individual entry looking for the error or unrecognized items. Future technology or software programming may alleviate this problem, but for now, be aware that unless every problem is identified, GIS programs will have some drawbacks.

If the incidents in question are once again only brush and dumpster fires that only the fire department responded to alone, limited information and data would have been reported. You will have to combine the methods used above and in chapter 6 to identify patterns. Choose the incidents, enter them into a separate search data group, place them on a map, circle the area, and back search previous fire activity for that community. Compare this activity with communities of similar demographics and use the method of identifying similar traits to clarify the likelihood that any misspelled and improperly entered incidents are or aren't the work of an a MFS. These techniques should help to clear the fog that will surround those incidents where no follow-up was done after fires were extinguished.

One of the easiest ways to start a false track is by including AFA (automatic fire alarms) in the analysis data. If these incidents are not designated on the fire reports or data and the structures in question have an ongoing faulty fire alarm problem, then the pattern will be greatly distorted. The designation must be made to differentiate the AFA from all other incidents. This is another good example of why follow up questions and cross-referencing must be conducted. Automatic alarms should be identified within the original source of data. If not, it will require the clerical work to correct the error and weed the AFAs out. It should not be that difficult, as the AFA types only pertain to structures. The reason for AFA incidents will vary. Systems may malfunction due to electrical shorts caused by frayed or loose wires creating a ground and faulty or dirty detectors can cause malfunctions. Accidental activation can also be caused by dust from cleaning or construction. Detectors can become dislodged. While someone is moving objects, an item can be dropped on a pull station. Whatever the cause, by identifying what type of structure is located at the address, the ability to quickly rule the incident in or out will be established. In most rural or suburban neighborhoods, this will be accomplished by identifying the address. A simple check to see what type of structure it is (for example, a structure in an all-residential neighborhood) will likely reveal the alarm to be faulty. This will not be as easy to identify in an urban setting. In urban conditions, check with the central dispatch or responding units such as fire companies and patrol units. Identify whether the system is residential or commercial; if it is a residential system, confirm if it is a multiple residence or single residence alarm system. The reason for this is that most single residences only have detectors whereas multiple residence alarm systems have detectors and pull stations in common areas. This would include hallways, staircases, and common entrances. If these are the locations for activation, then it can be the work of an MFS or someone who enjoys causing malicious false alarms. AFA incidents at commercial and industrial settings may be more difficult to identify. Nonetheless, check to see if any response units can identify the location and type of activation at the structure. On the other hand, if the incident occurs at a school or institutional

setting, it may not be a faulty system. The key here is not to confuse accidental or electronic malfunction alarms with MFAs (malicious false alarms, that is, the intentional pulling of fire alarms
boxes) in public facilities.

MFAs are easily identified in most modern fire alarm systems.
These alarm systems will usually note the time, date, type, and
location within the building of activation. This is all the information needed to start the process. Older alarm systems will require
more fieldwork. In large buildings, especially with public assem-
blies or institutional settings, the building maintenance staff will
usually be the best source for this information. Many building
maintenance staff keep records on drills and tests for insurance
purposes. Any irregularities in the system will be identified in their
records or logbook. As far as checking with alarm companies for
information on alarms, this can be a hit or miss approach. Alarm
companies vary dramatically in their operations, style and quality
of documentation. Check with other professionals who deal with
the alarm companies on a regular basis. These professions would
include fire prevention officers, fire inspectors, facility managers,
facility safety directors, and fire alarm testing agencies. MFAs may
be the first or original incidents in the career of an a MFS.
Therefore, be sure to follow up on MFA incidents that occur
repeatedly during a short period (two to three months). In educational facilities, check to see if MFAs coincide with midterms or
final tests. Incidents that occur during these timeframes should be
noted, but not added to the tracks unless the pattern of fire activity shows obvious signs of being the work of juveniles or malicious
mischief incidents.

MFA should also be considered if they begin to show traits
similar to the known suspicious incidents. An example of this
would be college campuses that are experiencing a rash of van-
dal fires. A check of the fire alarm records would be a good place
to start. Once the thrill of calling in or activating the fire alarms
wears off, the unsubs tend to escalate to the real thing. During
my research into MFS incidents I noted at least three college
campuses throughout the US that had suffered multiple fire incidents. A fourth located in the Northeast had a fatal arson fire dur-

ing the writing of this book. Although the details were limited it was reported that the dorm where the fire occurred had suffered eighteen MFA incidents during the previous semester. This particular case is a prime example of why more research and investigation into the MFA problem would be required before jumping to any conclusions. (To help in this analysis, use the methods explained in chapter 6.)

Rekindles are another issue for consideration when trying to prevent false tracks. Rekindles or re-ignitions usually occur due to deep seated fires within structures, dumpsters, piles of organic material, and below the surface at wildland or brush type fires. Be sure not to consider rekindles or re-ignitions. Rekindles in structures should be easily identified due to the fact that most structural rekindles occur within hours after the original incident. If a fire is noted at the same address within twelve hours, the chances of the fire being a rekindle are high. Check with the fire suppression crew and the fire investigator. Identify if the cause and origin for the fires are consistent with rekindle conditions. It is possible that the suppression personnel were not successful in totally extinguishing the fire. This would cause the fire to re-ignite hours after the units have departed the scene, especially in areas that were not properly overhauled due to dangerous conditions, such as fear of building collapse. If the origin of the fire is identified as a location where overhaul was not completed, for whatever reason, the fire can be classified as a rekindle. This is also usually the case with dumpster, refuse bins, garbage containers, or compactor fires as well. Unless the dumpster is completely emptied or flooded with water, the potential for a rekindle is high. Once again, time will be a key factor. A dumpster fire will rekindle within hours of the original fire. Because of this problem, be sure to identify if the second dumpster fire is in the same dumpster as the original fire or has the first dumpster been removed. Also, be sure to check that this is the only dumpster at the premises or if it is one of many at the shopping mall, construction site, or manufacturing sub-division. For instance, in a group of eight dumpsters, if a second fire occurs in the same dumpster within hours of the first fire, this would indicate a rekindle. When two or three fires occur in separate dumpsters at the same location, this would indicate something out of the ordinary.

The same cannot be said for deep-seated wildland, brush, and decomposing organic materials. These types of fires will be the most difficult to determine as the work of an MFS when they occur proximal to the original fire. MFSs, especially those in the initial stages of setting fires, will tend to ignite fires in the same general area. In most cases this will be easily identified as the work of the MFS. The only exception is when conditions occur that favor rekindles; the potential for deep-seated fires must be ruled out before proceeding. These deep-seated fires retain all the heat produced by the fire. Depending on the moisture content, the heat can be conducted to an entirely new unburned area. This may occur within hours or days later depending on the weather conditions. If your pattern search contains areas where these types of fires occur, it is strongly recommended that you speak with the responding fire suppression units to identify if the fire incidents were in the same growth area or a new growth area. The new growth area should have easily identifiable breaks between the vegetation that develops in each area. It is also recommended that you contact the fire investigator (if any) and, if possible, an expert on land management in that area who can identify these conditions. The experts would work for bureaus of land management from federal, state, and local parks as well as protected land councils. Another resource is the National Weather Service, that can identify when the conditions were right for a rekindle or deep-seated fire is the National Weather Service. Usually, these weather conditions require a substantial amount of time to develop and are limited to particular seasons. Experts in the area should be able to identify the timeframe and conditions required for deep-seated rekindles to occur. Technology may also be able to help identify deep-seated fires. The use of thermal imaging cameras can be used after suppression activity to confirm total extinguishment. The same techniques cannot be said for spot fires.

Weather, as mentioned with brush fires, will play a factor in fire activity. Weather will also be a factor in leading to false tracks if not properly identified. Dry weather with moderate to high winds could be the culprit for causing multiple brush fires and possible structure fires occurring all around the same area during a particular timeframe. Embers traveling over the winds can cause ignitions and spot fires miles from the original fire. Spot fires can also be created in wind-driven conditions during structural fires in and around dense-ly undeveloped or wildland areas. One of the greatest potentials for this type of occurrence is near the so-called wildland/urban inter-face. Structure fires in these locations can create spot fires that can grow rapidly when fanned by the winds that carry the original embers. Some secondary fires can grow larger than the original incident. Spot fires are usually caused by these burning embers, which are carried some distance from the original fire by a steady wind condition. This wind condition is usually when the wind is more than five miles per hour and has continued to blow since the original fire. In some cases the original fire is of such intensity that it creates its own thermal column that can then carry the embers a considerable distance from the fire scene. These embers have been known to ignite combustible or flammable materials that they come in contact with. A good example of this would be a housing project under construction. Even though housing projects are favorite targets of MFSs, that alone does not automatically mean that two additional fires in the same complex occurring in struc-tures of exposed wood frames at the same time is a probable MFS occurrence. In this case, you would need to factor in where the two homes are located–directly downwind from the original fire scene. Spot fires can be easily confirmed by checking with the weather experts in your area and by the cause and origin report. The cause and origin of all fire scenes should be checked in situations where spot fires are a potential.

Severe storm conditions could also cause false tracks. In certain parts of the country, thunderstorms have been known to circle or hover over particular areas before breaking up or moving on. During these circling thunder cells, multiple lightning strikes can occur in the same community. Structures hit by lightning have been known to develop into serious fires causing severe damage. In cer-tain sections of the US, large iron content in the ground has been considered as the cause of high concentration of lightening strikes. In many western states, lightning strikes account for the high per-centages of wildland fires, as compared to the northeast where strikes account for a very small percentage of wildland fires. The New Jersey State Department of Forestry has calculated that light-ening strikes account for less then 10% of the wildland fires while manmade fires account for more than 90%. Lightning has also been known to strike power lines and cause an occurrence similar to the back feeding of power to structures. This phenomenon can occur through grounding wires and can cause one or two structure fires simultaneously. I personally know of two or three homes that have been purchased from homeowners by the local power company after fires where this phenomenon had occurred. Basic data from fire activity will not identify lightning strikes; therefore, it is recommended that fire investigation reports be checked when two or more fires occur in one area within the same timeframe. A good fire report or field report should note the weather conditions during the incident. If this is not the case, other sources will be required.

A good source for weather information and conditions during simultaneous wildland fires or simultaneous structure fires is the National Weather Service. The National Weather Service keeps the historical record of weather conditions throughout the United States in their regional forecast centers. Weather records are also maintained by other federal agencies such as the US Fish and Wildlife Service, the National Park Services, the Bureau of Land Management, and the Environmental Protection Agency. All of these agencies keep weather information to varying degrees and for varying reasons. Check with these individual agencies in your area to see what they can offer. Some agencies may even keep records that are referred to as the Fire Weather Index. The Fire Weather Index is based on weather collection stations linked to a computer with software designed by weather experts to analyze current conditions and document historical trends related to burn potential or fire load of wildland growth. This index is designed to identify how weather conditions will effect fire conditions in undeveloped wildlands. The index is useful in planning suppression tactics and resource deployment for a particular day and future operations based on forecasts. The index can also be used by investigators and law enforcement for establishing additional patrols, surveillance, and denying access. This will be covered in-depth in chapter 12. As for this chapter, the index is important in helping to avoid false tracks because it will identify conditions that will cause a higher probability of rekindles, spot fires, natural or spontaneous ignitions, and accidental manmade fires. Weather information can also be gathered via the Internet or through public and private local forecast stations. Historical records of weather conditions for a region can also be gathered from private weather services, which may charge a fee for the data. These services also offer to help locate and identify lightning strikes. Lightning, as discussed earlier, can be quickly ruled out or supported based on the information from these services.

The issue of spree fires should also be considered when trying to avoid false tracks. As covered in chapter 2, the reasons for spree fires are usually not consistent with MFSs. Spree incidents will cover a specific timeframe, usually preceded by some type of event or incident, which encourages the fire setting activity. Spree incidents can also precede or accompany other acts of vandalism and violence. Some mob mentality groups will set fires to divert attention from their looting activity. This is not to say that the information on spree activity should be discarded. Spree fires may be the breeding ground for future MFSs or those prone to other acts of violence, especially malicious mischief or the activities of the power-oriented psychopath. Based on this potential, spree incidents should be logged and filed for identification in preparation of future occurrences. These sprees may have their own patterns, but rarely are they part of the primary track.

Because sprees may led to future incidents, it would be useful to also keep a separate log or data base by communities, fire departments, and dates in a book or computer file for future reference. Keep in mind that the fastest way to discern spree fires from an a MFS is the time distance technique. When researching the fire activity, check to see if an a MFS can travel from one fire incident to another within the timeframe of the reported incidents. In a spree situation, many incidents are usually reported simultaneously. Do not confuse this with multiple fires in one day, which occur a few hours apart of one another in varying locations. An MFS will set one fire and then move to another location attempting to ignite additional fires.

When a series of fires appears, for all practical purposes, to be spree fires there is still an outside chance that an a MFS is at work. John Orr (the MFS and Fire Investigator) traveled to communities all over northern California attempting to set multiple fires at the same time. The fires would take time to ignite and propagate. In some cases, the second and third fires would self extinguish. This individual was an accomplished fire investigator who knew how to set fires. As a fire setter, he was an organized serial arsonist. However, even with his knowledge and experience, his success at timing multiple incidents simultaneously was limited. During his trial he stated that only 50 percent of fire ignition devices are successful. This is not to say that it cannot be done, especially if more than one individual is involved, but it would be extremely difficult to accomplish. It would also require some form of delay device to get the fires to occur simultaneously consistently. Unless the targets are similar with similar fire loads and similar ventilation, the chances of uniform fire growth and simultaneous success are low. Therefore, the traits that we use to identify spree fires from MFSs are time and distance from the incidents (or duration between the incidents) and what events or incidents preceded the fire activity.

An obvious contrast between spree fires and MFS fires occurred in 1999 in the states of Tennessee and Louisiana. The subjects in Tennessee (Cookville) set approximately 100 brush fires in four counties during one night. The fires were all along the roadway and occurred in a timeframe of one eight–hour period. Although some of the fires burned simultaneously, none were ignited or reported simultaneously. The distance between fires was easily traveled by automobile, but would have been difficult, if not impossible, by any other method of travel. The fires were relatively small (approximately the length of one football field) and easily extinguished. The subjects in Louisiana (Baton Rouge) set approximately 24 fires in abandoned buildings occurring also along one road. However, the fires occurred in groups of five a night. The distance between the points of ignitions was determined to be within a bike or automobile ride of one another each night, with none being reported simultaneously. These incidents occurred over a five-month period. In the case of the Tennessee fires, a spree was occurring. All one hundred of these fires fell within a time period of hours along a specific distance consistent with the definition of spree fires defined in chapter 2. In the Louisiana example, the 24 incidents occurred over a timeframe of five months. There were cooling-off periods between each 24 hours of fire events. As for the distance, the events occurred over a distance of a few miles thereby reflecting a comfort zone typical of a MFS. This scenario is consistent with the definition given in chapter 2, for serial arsonists and therefore MFSs.

This brings us to another issue of false tracks, the subject of common denominators. Fires will have a common denominator such as occurrence off of a highway, within a fire district's borders, off a fire trail, or within the radius of a bike ride. We have to remember that fires do happen. Once again, remember to not connect the dots into a pattern that is not there. Two separate car fires during rush hour on the same road on two separate days may show a trend, but it does not make a pattern, even if the car fires occur within the active fire area. Now, if there are frequent reports of major events involving car fires and motor vehicle accidents along the same highway during rush hour, and all the reports have been false or unfounded, then there is something worth looking into. The same holds true for numerous incidents involving automobiles in a particular area or location. This may be a dumping ground for stolen autos or a case of arson-for-profit scams. Although numerous, it will be a rare case that this type of activitiy is the work of an individual or gang of MFSs. Check with the local fire investigator or fire department to confirm or disprove the area as a known dumping ground. If a community in question is being researched for MFS activity, the specific area of auto fire activity should be thoroughly investigated before being considered as part of the track or should be researched as a separate track. In the rare case where activity continues until there is a trend and subsequent pattern in the dumping ground, you should increase surveillance beyond its normal levels. The same potential trend may exist for a rash of dumpster fire incidents. In the case where dumpsters appear to be the exclusive targets, try to identify if this is occurring due to a carting war (disputes between the garbage contractors) or the renters of the containers are attempting to reduce the quantity of the garbage in the containers. This is sometimes done to reduce tipping fees since some carting companies will charge per garbage pick-up or additional fees for garbage beyond the normally scheduled pick-up.

When analyzing data reports, do not confuse multiple fires on one or two particular days as excessive. For example, five calls on one Wednesday in April and another five calls on one Wednesday in May should only be considered as two days of activity that occur on Wednesday. These two days may have had higher activity due to the weather conditions as discussed earlier. An example would be days that are particularly dry causing rekindles or spot fires. When conducting a data search by TIME and DAY as the key fields, each incident will be identified as separate events. This will cause the incidents to be counted separately allowing Wednesday to be identified as one of the most active days of the week for fire events to occur. If we use the example given and add to it that there have been seven suspicious fires on seven separate Fridays during the same period, remembering that the Wednesday incidents are not identified as occurring on two specific days, the individual incidents will skew the track. Wednesdays will be regarded as the most probable day for fire activity and Friday will be considered second or possibly overlooked. The result will then be inaccurate, causing assets to be misdirected from the higher probability day and target. To help avoid this situation when setting up the data, keep the DATE column close to the TIME and/or DAY column. This will help eliminate confusion in identifying patterns.

When analysis shows a pattern, be sure to recheck the data for errors in all the key fields. The last thing you want is to base an entire pattern on inaccurate information, which brings us to another one of the easiest ways for errors to occur. This is data entry whether collating by hand or by computer. When entering numerous incidents (say fifty or more) do not try to do so in one sitting. If the information must be entered in the database rapidly or if the process is being conducted by hand, you should take regular breaks. Once completing the data entry, another break should be taken. For those who are a hard sell on this idea, "PUT THE PEN DOWN AND STEP AWAY FROM THE DATA." Take five minutes, especially when working with a computer. Otherwise it will be easy to become fatigued and information will tend to blur or blend together.

Another pitfall would be to get caught up in the "chase". This is especially true when the data comes from one source and then another source (usually of the human intelligence type) is trying to convince you that an a MFS is responsible for fires in the area. Do not give in to the conspiracy theory. Work the facts, gather the information, weed out the unnecessary information, enter the refined information, analyze it, weed it out some more, and plot the findings. Do not skip steps especially when someone said something without the facts or evidence to back it up. This is especially true when dealing with CIs (confidential informants). Remember that they will tend to tell you what suits their needs whether it is true or not and whether or not it is the whole story. So, say for instance, you are going to rely on convicted felons for information; remember that they will only tell you what will get them what they want or off the list of suspects. Therefore, always try to hold back a little of what information is known or uncovered for use during future discussions with the CI. Bad information will lead to the formation of incorrect patterns and analysis. Simply put, bad information in equals bad information out.

To help avoid false tracks, remember that it is important to stress quality, not quantity. If the incident list contains 50 possible or potential suspicious fires, but only 20 can be verified, analyze only the 20 verified incidents. Then work to confirm the other 30 incidents by establishing the consistencies in traits discussed earlier and the methods discussed in chapter 6. Do not include them in the analysis unless the consistencies are first established. Keep in mind that dumpster, brush, or wildland fires will be difficult to determine unless a cause and origin investigation has been conducted. As stated, these types of incidents usually go unnoticed by investigators, especially if they are incidents that have occurred months ago. If they cannot be verified as suspicious, try to at the least identify if they are manmade or not.

For example, whether it was a carbon build-up from a diesel engine or an unextinguished cigarette, manmade fires were all created by human activity. For example, a poor running locomotive may have caused numerous wildland fires that occurred near railroad tracks. Instances such as traversing up a steep grade or down hill will stress the mechanics of the locomotive, causing the diesel engine to run inefficiently. This inefficient combustion will cause the engine to spew out hot carbon particles. The downhill motion may also cause the brakes to overheat and spew off hot metal fragments. This is just one example of a fire caused by manmade products operating in that specific area.

When there are no signs of criminal activity, this information will still be useful to fire prevention officers. Remember that their primary goal is to prevent future occurrences whether they are suspicious or not. Prevention and environmental enforcement officers are concerned about things such as equipment that is operated improperly and poor control of campfires. It may be that after investigation it is discovered that the fires are suspicious in nature.

For example, the potential to have six carbon particles or three catalytic converter parts and three unextinguished cigarettes in the same area in a two or three month period may be highly unlikely. These facts should be confirmed. A good course of action would be to discuss the likelihood of their related occurrences with fire wardens, park rangers, and fire prevention officers. If necessary, contact a wildfire cause and origin investigator and search the area for evidence.

This brings us to one of the final points on false tracks–analyzing data from unfamiliar or unknown terrain. Always try to get a feel for the land, even in densely populated or urban areas. Failing to do this can lead to mistakes, especially when trying to determine the travel time of an unsub, observation points, and travel routes. For example, do the travel routes taken by the unsub go through a particular ethnic or socio-economic area where someone from other than that community would standout? There are ways to help overcome these problems, some of which are discussed in chapter 8. Unfortunately, these solutions can be costly and have limited success. In many instances, the best solution is to visit the active fire areas to truly get the lay of the land and understand the people who make up the community.

So, let us say that something stands out in the research that leads you to the conclusion that the fire activity has signs of a pattern. First, check the neighboring communities for fire activity. During this search, it may be noted that multiple communities have not had the same consistent traits in their fire activity. The next step is to see if fire activity occurred prior to the search timeframe currently being researched. If not already done, recheck with other resources. This may lead to establishing patterns of fire activity that began years ago, or it may show no similar traits. It is not perfect, but remember, nothing in this process is. It has been developed over years of trial and error. It will continue to evolve even more over time with more searches.

Chapter 8

HOW TO PLACE THE INFORMATION ON A MAP

Mapping and the development of maps are the most direct reference to location. But location is not the only benefit to mapping. It has been stated by the ESRI (Environmental Systems Research Institute) of Redlands, CA, that an estimated 75% to 85% of all the information collected by the government could be placed onto a map. Examples of such information include the names of property owners, school locations, census reports, utility access roads, and much more. New mapping technology will undoubtedly help make this a reality. This is important because mapping helps to explain the means by which MFSs operate. For example, not only does mapping software allow the user to identify the locations of fire activity, it also allows the routes taken by a MFS to be identified and monitored for new incidents. Once the primary location (ground zero) is found, other locations can be checked for characteristics and trends that coincide with the primary active area. You can access government networks through the GIS (Geographic Information Systems) and obtain valuable information that helps identify trends and patterns and that can be analyzed directly on the map in record time. This vastly improves your ability to quickly identify the routes taken by the MFS and the locations of all his or her activity and will eventually lead to the identification of the suspects.

Suffice to say for now that even without the latest technology, location, as discussed earlier, reveals a great deal. It can help to identify the type of offender and reveal clues as to motivation. Incident locations will yield the forensic evidence needed to link the MFS to the activity. In most fire incidents, victims will be linked to the location, having a home, property, or both that were located in the target area of the MFS. In most incidents, the only damage will be to the property at the location of the fire. This is important not only for forensics but also for many investigators. When the victim tells his or her story, it reveals information about the suspect. Based on this information, victimoligists and investigators will need to know and see the locations. For example, they will need to ascertain what the land was used for and who had knowledge of it or access to it. Finding some reference point will help expedite their work. Remember that it will be the rare case that the same investigator is called to all the incidents. Most importantly, when patterns have been identified by the database search groups, the final stage is to allow profiling specialists to confirm the analysis with reference points for geospatial analysis, so that they can be used to further the tracking process and move to the next level, that of apprehension.

The idea of placing dots, or plotting, on a map is nothing new. In fact this is nothing more than the modern version of the pushpin map. However, the differences between this modern map and the old pushpin map are many. The list below describes these differences.

1. Pushpin maps were time consuming since everything had to be done by hand. Once the pins were placed they could be moved, but this would usually leave marks so one had to be sure and place the pin the correct spot the first time. If the maps were small, accuracy was reduced and it required a magnifying glass to read. Because of this fact, plotting would leave you at the mercy of the mapmaker.
2. A pushpin map was not reusable. All the pins would have to be removed. Removing the pushpins could be as time consuming as putting them in.
3. Saving the maps (especially if the case remains open) required a great deal of wall space, not to mention a lot of corkboard and many extra pins for the next case.
4. Pushpin maps only showed one thing: the location of the fire incidents. If the map maker was really ambitious, a leg-

end with different color pins designating the type of fires or the timeframe of events could be used. However you would have to choose which additional information you wanted. Under the old pushpin version, more than one item was not possible. Information beyond this would require you to use a new map.

5. A pushpin map does not lend itself to layering one map over another, better known as overlays. Breaking maps down by incident types, day, or time and then overlaying one on top of another would be difficult, if not impossible.

6. The pushpin map could only be shown in one location at a time; to duplicate it would have taken twice the time, not to mention trying to transport a wall map to all the different agencies that will need to view it and be briefed. Instead, it was required that personnel from all these agencies come to the map. This creates a problem when shift personnel, multiple teams, squads, and multiple agencies are involved. This can also be an issue when attempting to present the map at trial. If the map does not explain specifics, then the defense can use it to argue that the map shows too many variables to be considered, leaving open a question of doubt.

7. If the map was too small, the pins were so close together that nothing was decipherable. Supervisors may have realized that there were a substantial number of incidents to look at, but were not able to identify where they occurred or what they were. Enlarging the map led to more difficulties than advantages if not done properly. Usually, when copying a map, it could only enlarge one size; otherwise it was not legible. Legends, direction, and scales all became lost or distorted.

The pushpin map served its purpose and some of the principles involved with it can still be used to develop its handwritten counterpart. Unfortunately, this is done at the cost of time and with limited success. However if this is all you can afford to work with, then do so because maps will still have a greater advantage over data reports, lists, and simple pie charts.

Remember that the advantage of using maps whether they are done by hand or computer-generated is that you can view the whole picture and are thereby able to see the incidents and correlate all the pertinent information regarding each incident without having to be some spatial analysis expert. If you have access to a good mapmaker or cartographer, accomplishing this understanding with handwritten maps can be just as useful as the computerized version.

Distribution of handwritten maps can be improved with some simple modifications. For example, a regional map or fire district map used to locate the cluster area can also be duplicated with the section of interest copied to reproduce the active fire area. Make additional copies of this map and put them aside. Mark the fire locations with a colored pen or highlighter. We will cover how to color code the locations later, but it is easy to see how additional maps could be reproduced and then highlighted. The copies that were put aside can be used to produce more specialized maps such as particular days, times, or months. The use of a copy machine will greatly increase production efforts while decreasing production time. Having said that, the chances of the handwritten production process competing equally with a computer-mapping program are minimal at best. Nonetheless, remember that any map is better than no map.

Of course the entire map is not handwritten. Use a pre-drawn map such as those mentioned: police, fire, or school district maps. Maps found in the local or regional phone books may also work. Almost all parts of the country have some form of road atlases that can be purchased. Up to the mid-1990s the jurisdiction I worked for relied on maps from our central dispatch center to verify locations and identify fire and police sectors. It may be possible to obtain used versions from public safety communication bureaus. Many school and fire districts will supply maps for a nominal fee or for free. If you have a computer and are online, maps are also available on the Internet. These maps can usually be downloaded for specific areas. Some state agencies will have Internet sites with regional maps available. Unfortunately, if a computer is not available for mapping, the likelihood of it being available for this use will be slim at best.

Whatever type of map is chosen for the process, keep the original and make additional copies to draw on. Based on this requirement, the use of a copy machine will be necessary. If a copy machine is not available or your agency currently uses a single print copier with no zoom or multifeed capabilities, it will be necessary to obtain or gain access to a more advanced machine. Check with your agency's local government to see where they have their copies or printing completed. Many jurisdictions have their own print shops or contract with a local printing company. Print shops or print services usually have the most current copy equipment. Color copies are not necessary, but if offered, would help. If the use of a copier is not possible, or only the basic copier is available, do not give up. Try using transparent paper or overheads to draw up specific maps. This will not be easy and will require some technique to line up the map and transparency properly, but it can be done. Remember, maps will be one of the quickest references for others to identify and understand the presence of a MFS. Many times they will be critical to convincing others to take action. Hence, they are well worth the extra effort.

We will spend the remainder of this chapter discussing computer mapping techniques and features. The modern pushpin map is accomplished with computer mapping software that will find the location and allow for the plotting of fire activity with customized information. For example, current software is able to plot a specific symbol for that type of incident and a specific color to designate that particular day of the week or time. Other software programs allow for the placement of a flag with the date, time, and type of incident in it. This becomes very helpful when plotting multiple incidents in the same area. Mapping software varies in capabilities and price. Remember that the most expensive does not always translate to the best product. This is especially true when it comes to computer software and specifically computer mapping software. Once again it is recommended to check with your municipality to see if they are currently using a mapping software product or GIS software. If this is the case, be sure that equal or unencumbered access to the software is available. This may be asking a lot in some jurisdictions, but without that access, little will be accomplished. If there is no access or no system currently exists, then it will be necessary to find a system. When deciding on a mapping software program, keep a few things in mind. First, be sure the system is user friendly. Some GIS systems in particular require hours of training. This training can be cost prohibitive as well as restricted to the primary users. One other disadvantage to consider would be that if you establish the GIS program first, it will require that you enter all the data. This will mean spending a good deal of time transferring and entering data for your work as well as for others. Once again, check with GIS coordinators in other jurisdictions.

In the municipality where I am employed we have two separate GIS systems–one for public safety and another for everyone else. This in my opinion was not a good idea do to the fact that it not only restricts the flow of information, but also increases the cost and limits the available training. It also brings the synergistic effect that grows out of people working together to a screeching halt. It has also requires that the input of information be duplicated for almost all the data. Therefore based on my limited experience I am willing to offer this opinion on future GIS purchasing. Even if the system is not perfect for all the needs of a particular agency within the jurisdiction, remember many minds working on the problem will usually lend toward finding a better solution. Not to mention the fact that future versions of the software will more then likely resolve the needs.

Off-the-shelf mapping software will cost anywhere from as low as $9.99 on the discount rack to as much as you want to spend. Some GIS systems and most off-the-shelf mapping programs have tutorials or easy to use instruction manuals. Although this was mentioned earlier, the following is worth repeating: Cost of software does not always follow the old saying that you get what you pay for. Sometimes less expensive software will work better for your needs. For example, some software systems start from the beginning, developing their own "language" and code which costs money to develop and does not guarantee that it will work any better than something you buy off-the-shelf or mail order from some retail company. When researching the mapping software programs for the tracking process, look for the following:

- Compatibility with the current computer system(s) used
- Software at discount prices for older machines
- KISS (Keep It Simple Stupid). If the sales person or the label on the box recommends that you attend a 6-hour training lesson, then this software is not for you.

Within a few hours of loading and working the software you should be able to get started. As users continue to work with the software, analysis should become more efficient.

Many of the mapping programs on the market are based on vector and raster digitizing. Without getting bogged down in terminology and programming, these are the types of conversions required to store mapping data. Many GIS programs rely on digitizing data to accomplish the mapping tasks. This digitizing in simple terms is assigning coordinates along the X- and Y-axes of the image. One of the easiest ways to identify these coordinates is by using global latitude and longitude positions. This feature offers the advantage of highly specific, if not pinpoint, accuracy of all points worldwide. For the purpose of the tracking process, as well as other mapping processes, this data can be easily identified and documented using the GPS (Department of Defense's Global Positioning System). The GPS allows for, as nearly as is possible, dead reckoning. However, generally speaking the exact location of the incident is not required, especially when you zoom out to a wide view (say more than 1000' out). This is based on the idea that when zooming out, the detail will decrease and incidents may become blocked or hidden behind one another. This may be a common occurrence in suburban and urban areas with multiple incidents occurring within a city block. Zooming out to show the big picture may hide much of the flagged information. Therefore, it may be required to move the flags off the specific location to a more general position, as close to the incident location as will suffice when looking at the big picture. Other types of incidents may require more specific locations or smaller more detailed maps. For this purpose, the accuracy will be an improvement over the original pushpin method. An example would be forest and wildland fires, which may have a cluster zone that spans three miles in circumference and will require somewhat more accuracy for crime scene analysis than a city block.

At this point it may appear as though the cost of this process is escalating. The thought of being over budget is probably looming around in your head. Do not worry; the items we are discussing are not that expensive. These items are discussed in chapter 11 and will easily fit in a budget. Having said that, understand that an accurate user friendly GPS can be purchased (at the time of this writing) for approximately $150.00. Some GPS units will cost more and some less. Keep in mind that less than five years after the introduction of personal GPS units, the price tag dropped $500.00. As these units become more popular, the price will continue to decrease. Software that will accomplish the goals of pinpointing the origin had been mentioned in chapter 3. This software will not only identify the latitude and longitude, but for an additional few dollars it can load the data to and from GPS units. The cost of this software is approximately $35.00. Some companies will sell the complete kit, which includes the map, software, data cables, and GPS unit for approximately $200.00.

Features needed in Mapping Software

As far as specific needs and functions of the software, the following features should be in the software. These features are listed in the order required to conduct a pattern search. Once again, this is not the only way to conduct a search pattern. Whatever software is chosen, it should closely mirror or offer the following features:

- **The software should be able to find and plot locations.** This feature is key because it will save many hours in the process. Many programs will only narrow the search down to addresses with the similar spelling. This will speed things up, but not as well as other software search systems that identify the street address and display it on the map. Be sure that the software will distinguish between different towns and zip codes. This is an important specification that will increase the accuracy and decrease the search time. Some programs will break streets down into blocks based upon the street number. These are not always accurate, so check the accuracy using known addresses.

- **The software should be able to post information about the incident next to the plotted location.** Although this feature adds time to the plotting, it will enhance the map to levels unknown prior to this development. This ability will also help decrease confusion among other viewers when it comes to all aspects of the plotted incidents. This information includes, but is not limited to, the date, time, and type of incident. The day of the week can also be included if it is necessary or will help viewers of the completed map. This feature will allow for quick reference when comparing the similarities to incidents, not only by location, but also time, date, and type. Use abbreviations for the type of incident and day of the week. For example, a structure fire that occurred on January 2nd of 2000 (a Sunday) at 23:30 should read as "1/2/00 23:30 Sun, Struct." on the flag.

- **The software should be able to display geographic locations or landmarks and a scale for measuring distance.** Examples of geographic locations would be trails in parks, school buildings, and town or community boundaries. Some of these features may already exist in the software, but if they do not, they need to be added.
- **The software should allow the user to write information directly on the map.** This makes it possible to incorporate the map meta data, which would include details of what the map represents, who made it, when it was made, and any additional information that would be required for the reader. An example of this would be an orientation pointer or compass showing the four headings (North, South, East, and West). The ability to add dotted lines, thin colored lines, colored circles, squares, or other shapes will be very useful in detailing the size of the fire area and the proximity to important locations such as bus routes, fire district boundaries, and utility access roads. This ability to improve the current map will convince doubters such as employers, coworkers, teachers, fellow students, neighbors, and family members of the suspect's involvement. It will also allow viewers to comprehend better when new roads or structures are developed after the map was originally designed.
- **The software should have the capability to save all the information you have developed.** The last thing you want is to go through all that effort with nothing to show for it. This is important for the updating and the deleting of inaccurate information on the maps created.
- **The software must be able to print legible maps (preferable both in color and gray scale).** If the maps are only available in black and white, they must show enough detail when printed to distinguish specific details such as borders

and other demarcation lines. These distinctions must be available whether the map is printed on a state of the art laser printer or a noisy, old single-ribbon dot matrix unit.

The key functions discussed here are what will make the system work. Other features, which will help enhance the map but are not required, should be considered with a cost-to-need ratio. Simply put, is the cost of this feature worth the added expense at this time? Do not automatically rule anything out; in fact, all of these features should be open to consideration when purchasing software. One of the more helpful features to consider is the measuring of distances and areas. This will be very helpful when identifying the size or area of the cluster zones and the time verses distances traveled between incidents. Examples of applying this feature would be calculating the time it takes to travel from one incident to another within a few hours or in one day. This will also help to rule out incidents, by establishing the operational area of the MFS and the location of cluster zones. Travel time can be estimated and established for varying methods of transport. Once a fire is reported, surveillance personnel will know the estimated time of arrival (ETA) for an unsub to travel to and from the cluster zone. Subjects moving in and out of the area can be interviewed for information. As mentioned in chapter 7, once these skills are mastered with a few keystrokes or movements of a mouse, the time and distance can be used to help rule in or out subjects possibly within seconds.

Some mapping programs can also use photos obtained by satellites. This is where another useful aspect of mapping comes in–topography and aerial photography overlays. The topography feature is also helpful when establishing time-to-distance calculations when hills, valleys, and wetlands are involved. This can be a concern if the personnel conducting the analysis are unfamiliar with the land. Unseen natural barriers, elevations, and/or land depressions will not show up on two-dimensional maps. These land features can also affect burn patterns and burn activity. This brings us to another helpful map feature known as the aerial photo overlay. These photos are purchased through a number of private services that are available locally, regionally, nationally and internationally. The quality will vary with the service. Most are taken from aircraft at varying altitudes at what is known as 4-meter resolution (this refers to quality of the photo based on the fact that objects up to four meters in size can be identified from varying altitudes). I sometimes question this concept, but if you are or know an aerial photo interpreter I guess it is possible to make out items four meters in size. Understand that the clarity of these photos is not like the ones taken with a 35-millimeter personal camera. The greater the zoom, the grainier the photo (this means that the quality of the photo will be reduced). Even though it will be possible to make out trails and tracks, it will probably require an expert to tell you what made them. If you want to use this technology, be sure that the photo data is compatible with the software you will use. In most cases, aerial photography will not work with less expensive software. This does not mean that aerial photography should be ruled out. It simply means you may have to spend a little more for the software. The software we use (Precision Mapping) will work with the photography, but requires upgrading and technical support (See Map 8-1).

STATE RT 50

TREE LANE

MAJORITY OF FIRE ACTIVITY

PINE CT

XYZ Fire Activity
For 2005
Prepared by
C. Cinder
4/5/05
N

0.5 miles

DIRT ROAD

Map 8-1:

Aerial photograph with map overlay depicting 4-meter resolution. (New York State GIS Clearinghouse, "LI South Shore Project, www.nysgis.state.ny.us/raster.htm)

The newer generations of photos now available are said to have two-meter or higher resolution than that of previously available images. Photos and maps are slowly becoming available through the Department of Defense's National Photographic Reconnaissance Office and also from the former Soviet Union, the latter being available on the Internet. Currently, most of these photos are of what was considered to be strategic locations. Some of these locations did include some urban areas. The process of aerial photographing, digitizing, and geocoding to computer images that match up with the maps is expensive. Usually, this process is undertaken by more then one agency to offset the cost. Check with local and state government agencies as well as utility companies in the region. Check with agencies that do use the imagery to see if they can supply you with printed photos of the area in question. These photos may also be available through local, state, and federal highway departments. They have been known to regularly photograph roadways and surrounding areas for future expansion projects. These will be helpful in identifying natural or manmade geographic features not noted on the computer mapping software.

Now that we have discussed the items and techniques required we can continue with how to apply these requirements. The first step is to identify all of the incident locations. If the incident can be labeled on the map, do so with the information discussed earlier in this chapter. This will be difficult in the handwritten version; some form of color-coding is recommended. As far as color coding of hand maps is concerned, write the date and time in circles near or on top of the incident locales on the map chosen. Place all of the incidents on the map based on your pattern search analysis. At this point, make copies or have transparencies available (at least 3) and put one aside. Once again, break out the highlighters and colored pens and use them to identify types of incidents. For example, highlight car fires in yellow or go over the circle with a red pen. Use a green pen or highlighter to circle brush fires, blue for structures, and leave the dumpsters in black (See Map 8-2).

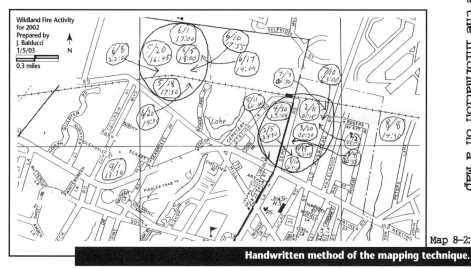

Map 8–2:
Handwritten method of the mapping technique.

This same technique can be used for days. For example, Sundays in blue, Mondays in red, Tuesdays in green, Wednesdays in yellow, and so on. Time can also be color coded to make it stand out better or if it will not fit in the circle. This is helpful when dealing with many incidents on a map. For example, incidents that occur from 5:00a.m. to 11:30a.m. could be colored red and identified as morning incidents. Incidents from 11:30a.m. to 5:00p.m. could be colored green to designate afternoon. Incidents from 5:00p.m. to 11:30p.m. could be colored blue to designate evening. Finally, incidents from 11:30p.m. to 5:00a.m. could be colored black for overnight. As for the computer version, color-coding can also be employed, although simply abbreviating key information, as mentioned earlier on the flag (or tag) will work just as well. Colors or symbols can be employed (if available in the software) to draw attention to specific days, times, or types of incidents. Unfortunately, this can sometimes limit the information or create a bias toward certain incidents. For example, by placing a flag with each piece of information, all the map-readers will have the pertinent information visible to them without having to refer to any legend or explanation of details. By highlighting certain incidents rather than all the incidents noted, the map-readers eyes will be drawn to those specific incidents highlighted. This may influence the reader in a way that was not intended. Whether utilizing the hand or computer version, when placing the data on the map, try to use a standardized listing or order. Use whatever order is convenient and legible for the needs of the tracking process. The following is the recommended order: date, time, day, and then the type. When drawing lines, squares, circles, or shapes, try to also be consistent with color, thickness, and type. For example, when designating district borders, use one color or dotted lines throughout. Use a different color, different shape, and different thickness to designate the cluster zones or geographic hunting ground.

The first map developed will be a plot of all known and/or probable incidents. Start with these incidents before adding any possible and prior uninvestigated incidents. This first plotted map can be thought of as a baseline to help judge unsubstantiated information. Use the "DATE & TIME" search group to accomplish this task. This map will be easily added to because the data group information develops in chronological order. When producing the map by hand, it is recommended to first use a pencil until the incidents are verified. Using a pencil is also recommended if additional former incidents (unknown during the initial plot) are added to the original list.

Once all known and/or probable incidents are plotted, the next step can begin. This is the task of identifying cluster zones. At this point it may be required to enlarge the maps in and around the cluster zone or zones. Use the computer-mapping program, or if by hand, any other maps that are available, to identify specific features mentioned earlier in this chapter about the cluster zone area. These items include demographics, topography, and the size of the cluster zone. Try to establish any similarities in location or in the types of activity within the cluster. Some examples would be incidents near parks or schools where wildland fires occur. Other items include factories and vacant or abandoned properties. This can also be an opportunity to plot the distances between incidents and the size of the cluster zone area. Aerial photography can also be employed to identify geographic and structural features in and around the cluster zone. Examples would include clearings, access roads, and vantage points to view fire activity (see photo 8-1).

Walter's School

Photo 8-1:

Wooded area surrounding residences and a school in a suburban neighborhood showing roads, trails, and paths. (New York State GIS Clearinghouse, "LI South Shore Project, www.nysgis.state.ny.us/raster.htm)

Save this information as it is identified. Once the plotting is completed, if no clusters are identified, it may be necessary to re-evaluate the source data before proceeding any further. Clusters should be one of the key components to tracking MFS activity. The lack of this component should cause concern, but not rule the pos-sibility of MFS activity out. Items to check for would be the types of fire activity as related to the location. Remember that extremist groups usually spend a great deal of time and effort in choosing a target. For example, on one extremist website (the Earth Liberation Front or ELF) instructions and techniques in surveillance, spotting, and counter-surveillance are discussed. Those targets are often a considerable distance apart from one another. If the fire activity were geared toward known targets of extremists (government facilities or other institutions) this would be one of the considerations during the re-evaluation. The first or original plotted map should be the busiest looking of all of those created, due to the fact that this particular map will display all of the relevant incidents and data. The maps that follow will be based on much of the information that is extracted from this map, but not all, and should therefore not appear as busy as the original map. (See Maps 8-3 and 8-4).

Map 8–3:

Quantity Map—shows all the fire incidents analyzed. (New York State GIS Clearinghouse, "LI South Shore Project, www.nysgis.state.ny.us/raster.htm)

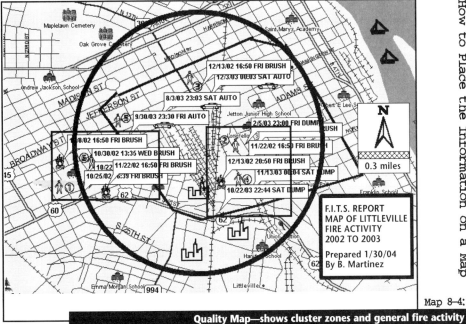

Map 8-4:

Quality Map—shows cluster zones and general fire activity

The second step is to develop a map based on plotting the activity by day, type, and time.

Once again, when developing this map by hand, use a new copy or transparency. To develop this map, use the "DAY and TYPE" search group data. The DAY and TYPE search group will help identify specific activity or days of activity in the cluster zones. If there are any trends in regard to specific days, this map will be the most helpful to include in the final report (See Maps 8-5 and 8-6). A map plot based on a particular month or week of activity should also be considered. This map will assist the investigator with interview questions of subjects and witnesses in the active fire area. This map can also be used to link particular events or occasions to subjects in the active area. For example, suppose during federal holidays, an increase of activity is noted around a particular neighborhood during the hours of 12:00 to 17:00. All other activity has been noted near a school between the hours of 15:30 and 17:00. This is a simplistic example, but should show how a map can be used to record the actions in and around the active fire areas. To assist in the development of this type of map, use the calendar plot that will be discussed in chapter 9.

Although all of the search groups can be plotted on maps, this can be time consuming and may not be worth the return for the effort. That is not to say that all of the other types of maps have been thoroughly researched and studied. On the contrary, it has been my personal experience that the other types of maps that have been developed have not been as useful to this writer as those discussed. Those discussed have yielded the most benefits when compared to the amount of time it takes to develop and use them. However, it is recommended that the reader attempt as many map designs as are useful to the analysis and to the dissemination of information as is possible. See the appendices for references to further sources on mapping and plotting.

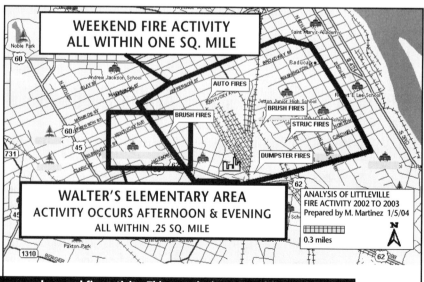

Map 8–5:

Cluster zones and general fire activity. This map depicts general fire activity that has been identified as MFS activity. This map shows quality. It should be considered good for a final report, but may not be as useful to analysis and tracking personnel. Note that it contains all the necessary information for distribution, including the author, date it was produced, what it represents, and the timeframe covered by the activity.

WEEKEND FIRE ACTIVITY
ALL WITHIN ONE SQ. MILE

12/3/03 00:03 SAT AUTO
8/3/03 23:03 SAT AUTO
9/30/03 23:30 FRI AUTO
2/5/03 23:00 FRI DUMP
11/13/03 00:04 SAT DUMP
10/22/03 22:44 SAT DUMP

F.I.T.S. REPORT MAP FOR LITTLEVILLE
FIRE ACTIVITY FROM 1/1/03 TO 12/31/03

DUMP – ROLL OFF DUMPSTERS
AUTO – VEHICLE FIRES

Prepared by T. Martinez 1/5/04

0.5 miles

N

Map 8–6:
Specific activity at specific times on specific days

How to Place the Information on a Map

Chapter 9

How to Create the Calendar Plot

Calendars can be extremely helpful once patterns and tracks are established. Calendars help to correlate patterns with other relationships or significant events. Calendars also help determine the unsub's schedule as well. Most people do not realize their own set routine. It has been said that humans are creatures of habit. The majority of these routines involve the basic needs of daily human life, those mundane items that are done on a religious schedule and are rarely departed from, except for some special or catastrophic reason. These things become so routine that they seem boring or meaningless. Examples include weekly grocery shopping, filling one's car with fuel, commuting, visiting family members, or visits to the doctor. Then there are those events that are anticipated and considered worth noting on an individual's social calendar. This kind of event may be the only bright spot in what was considered a dull day or boring week. Examples of such events may include cashing a paycheck, getting together with friends for a card game, going on vacation, shopping, and visiting favorite family members or friends. All of these events, boring or celebrated, can be identified with other events that have already been documented. For example, the purchasing of goods will, in most cases, involve a receipt. That receipt is a two-way record for the purchaser and the vender. Most modern purchases will involve a date and time stamped on the receipt. Some store convenience purchases and ATM money transactions will also be video taped. Even if no documented record exists of the event, some contact with others will occur providing an eyewitness account of location, time, date, and what occurred. Based on this information an individual's routine can be identified and plotted on a calendar.

Those same ideas hold true for the offender. The MFS routine may be linked to their daily or weekly routine. Details of this routine can be a wealth of information to the tracking process. It will help to establish the best days to interview the individuals that the suspect has had contact with, when to establish surveillance and what new maps to create. Calendars can assist in the map development by emphasizing time periods for map production, specific days of the week or specific weeks of activity to be plotted. Calendar plots may show the direct relationship of active time periods to new hunting grounds. This will help define the appropriate time periods for map generation. For example, the unsub may set three or more fires in a four-day period and then cool off for a five- to seven-day period. This may be a consistent trait observed in the fire setter's pattern and would be the most readily noted on a calendar plot.

Calendar plots may also be used in the reverse role; that is, calendars will help trace the incidents back to the original events or ground zero. Actually identifying ground zero is probably one of the greatest assets for the profiler or forensic psychologist when it comes to unlocking the mysteries of the unsub. Discovery of this site should not only help with identifying clues about the unsub, but also be used during the interviewing of the unsub. Evidence, odors, photos, and props from ground zero can help with the interview process. Unfortunately, ground zero is not always that easy to identify when multiple clusters are noted. That is the benefit of the calendar plot since it can help to identify the first or original incident. The information plotted may help with behavioral analysis and help identify an organized or disorganized MFS based on their schedule for fire activity. These are the details for the profiler, as well as the person doing the tracking to a lesser extent. Either way, it will not matter whether the tracker identifies the offender as an organized or disorganized type immediately. As routines and schedules are identified these can be used to establish if the offender is organized or disorganized. Routines are one of the reasons why firefighters, who turn arsonists, are so easy to catch. The fire setting activity can quickly and easily be matched against responders as well as the scheduled meetings, social events, and training of the Fire Department members. By inputting this information to a calendar of scheduled events, the process can be used to prove or disprove firefighter involvement.

To expound upon this situation further let me state that one of the reasons why many investigators first look toward the fire department to find the arsonist when multiple fire incidents occur is because routines are easily checked. Little effort is required to collect any and all evidence. In fact some investigators will simply confront the chief officers to see if they have taken on any new personnel. Based on the fact that records are usually kept on who attends the meetings, training, and (most importantly) fire calls the investigator can quickly gather enough evidence to confront the suspect with overwhelming circumstantial evidence. When investigating a volunteer firefighter, the simple presence of the subject at every suspicious fire and the fact that no fires occur during the same timeframe that the subject is attending meetings and drills will quickly narrow and focus the investigation. By identifying the times that a suspected fire department member leaves the fire station, signs off duty, or just prior to signing on duty, you will be able to establish a pattern that can convince any doubters that there is a problem.

A word warning at this point must be given. While writing this book a MFS case was developing in a densely populated community. Some of the investigators wanted to point the finger at the local fire department. I knew the fire chief in the community in question. He was a good man and would hand the unsub over bound and gagged if there was substantial or circumstantial evidence pointing to any specific individual within his ranks. But if there was no evidence that pointed to anyone in particular and it was just based on a hunch, the fire chief would take it as a personnel attack against him and his personnel. Based on this knowledge it was recommended that we first complete the process of tracking the fire activity in the area before confronting anyone. Fortunately all investigators agreed. The track was not consistent with the activity of a firefighter. In fact the track was not even consistent with one individual. The fire activity had occurred on both sides of the fire district borders and occurred during training and meeting times. The fires appeared to be the work of multiple individuals. The times, distances, locations, and types of the fire activity suggested juveniles. This was further reinforced with HI (human intelligence) that focused on teenage individuals on bikes. All this was done without involving the fire department or the chief who in the future would become a positive source of intelligence instead of negative force.

Another example of how routines can reveal information is in a situation where most fires occur during the middle of the week from the hours of 11:00a.m. to 11:00p.m. may suggest that the unsub has little responsibility at home or at work. Further analysis may identify that the unsub has no specific schedule and has been escalating in the type of fire activity he or she is involved in. You may observe that the incidents continue to occur in the same area, but at inconsistent hours of the day. These could be signs of a disorganized MFS, who has a great deal of free time on his or her hands and limited responsibilities with limited funds for transportation. The same unsub's calendar plot shows that no specific day is preferred, but the fires occur in two and three day clusters with multiple incidents occurring on those specified days. This would help reinforce the theory of an unorganized offender who is building up to more serious fire events. His or her motive is possibly based on frustration, anger, fear, lack of confidence, or any combination of all of those mentioned.

On the other hand, fires that occur on weekends between 11:00p.m. and 1:00a.m. may identify the only hours that the unsub is not responsible for someone or something at home or at work. With the weekend singled out and the time of activity specific, this offender may be conducting fire activity of a specific type although over a wide area. This could be a sign of an organized unsub using his or her limited time to do the greatest perceivable advantage. Another example would be fires that consistently occur on specific dates of every month or year. This would suggest an organized offender who could possibly be linked to anti-government, anti-establishment, or anarchist organizations. These are all examples based on non-scientific observation, but it can be seen how a trained behavioral expert could use the information to identify routines and form theories.

Once a subject is identified, the information that has been gathered from the calendar plot (along with the other data developed) can be used to validate the subject as a suspect. This can all be done without questioning or alerting the subject to any suspicion. Simply cross-reference the calendar dates and times with the suspect's work schedule, school attendance, organization affiliations, and/or social calendar as much as you possibly can to see if opportunity is consistent with activity. An example of this theory can be applied to the unsub who operated between 11:00p.m. and 1:00a.m. earlier in this chapter. It can be seen how the unsub's limited free time will quickly correlate with his or her schedule once the subject is identified.

Calendar plots also assist in assigning patrol requirements. Surveillance, as well as increased patrols, will be covered in greater detail in chapter 12. For the concerns of this chapter, it is important to realize how information relevant to surveillance can be gained. This would involve noting clusters of days that would be best for saturation and/or surveillance. Knowing what days of the month are more popular for fire activity than others will greatly assist in setting up patrols, surveillance, and questioning. For example, the absence of activity on every third Sunday or the identification of multiple incidents on every first Friday of the month will reveal something about a subject's schedule, possibly identifying the MFS's visits with family and friends on that particular Sunday. Fire activity on the first Friday may indicate that it is not a paycheck week for the unsub; therefore, he or she may not be able to go out and do the things desired. Therefore, the MFS may be looking for something else to do to satisfy his or her needs or to simply be entertained. In cases where intoxication is a factor in building confidence in the MFS to commit the act, the opposite may be true. For instance, an increase of fire activity on the second and fourth Fridays may help establish the subject's paydays because the MFS has the money to buy alcohol. In either case, this information can be used to check local bars and other entertainment spots surrounding the cluster zone to see who the regular big spender is or isn't on the active fire days. Interviews would also include persons who may have contact with MFSs such as businesses and establishments that surround the cluster zone. A recommended check of other criminal activity that coincides with the fire activity should also be considered. This additional criminal activity can also be plotted and, if necessary, tracked. Domestic violence, criminal mischief, or other disturbance calls are all examples to consider for plotting on the active fire dates. If during each month it is noted that specific weekends are targeted, that information would be the intelligence needed to authorize the surveillance.

Calendars can also assist in showing when the MFS is not active. A lack of activity can be as relevant to the track as increased fire activity. Examples would include if the person was out of town for travel, medical emergency, family events, and death of a friend. This was the case in Brooklyn, NY, from the end of 1999 to June of 2000 when a MFS set 18 fires in the community of Bushwick. This led to approximately 25 families being left homeless and the death of one child. The cluster zone was concentrated to within two blocks around the home of the MFS. Unfortunately, the pattern of fire activity stopped suddenly. Unknown to all investigating the fires was that the unsub was in and out of the psychiatric wards of local hospitals for the next three months before returning home. Upon his returning home, the unsub resumed his fire setting activity. Now the subject was setting multiple fires at multiple locations during each episode. He was finally caught on June 12th of 2000 when he was spotted at the third suspicious fire of that evening on the same street. When asked if he knew who set the fire, the subject responded "The Devil." When further pressed on who "The Devil" was he responded "I am."

As you can see from this example the gaps noted between fire activity can be used during suspect interviews. Instead of asking about the fires, ask about the inactive period. Something may have been going right for the suspect. The fires may signify what was going wrong (the so-called torched soul). Determine if the subject was away and if so, identify where the subject was. Check with the jurisdiction where the subject was for increased and/or unusual fire activity during their presence. Gaps in activity can also be attributed to the MFS having been scared off. For example, try to identify if a lack of wildland fire activity coincides with the opening of hunting season.

Calendars can also be used to look at activity based on environmental conditions. When thinking of the phrase environmental conditions, do not limit thoughts to weather only. It should also include atmospheric condition as well. Examples would include the phases of the moon, the solstices, meteorological, and astronomical trends. Some public safety veterans have strong beliefs that a full moon plays a role in human behavior. Whether it is true or not is unclear. What I do know is that when the issue is mentioned to an RN in the hospital emergency room, the patrolmen on the beat, or the firefighter on shift, they all dread working a Friday night shift with a full moon. I will not try to explain this phenomenon because I do not quite understand it. There have been many studies of these environmental effects on animals, plants, and humans by experts on the subjects. Some of these studies are referenced in David Icove's *Arson Pattern Recognition*. These studies are old (from the 1970s); nevertheless, the information should be considered.

I will now offer the following observations made by fellow investigators. First, a greater amount of activity is more likely to be noted when there is a full moon because the full moon is the identified point of reference. This is compared to other high volume or active days when there are no easily recognized reference points. The second observation is directly related to criminal activity. Based on the fact that nights with full moons and new moons are brighter, more crime is actually spotted by civilians and law enforcement personnel, not to mention that more people can attempt any number of things because it is that much brighter on that particular night. Third, the position or phases of the moon and sun affect tides and temperature. These scientific facts affect the activity of humans. Ships arrive and depart based on tides; communities flood based on unusually high tides. The amount of working daylight will increase and decrease based on the season (regardless of daylight savings time). These naturally occurring events will affect the individual schedule to some extent. Whether it is irritability due to long, hot, humid days or depression due to short, cold, damp days, followed by long dark nights, these naturally occurring events will affect certain individuals more than others within the same community.

Humans, whatever the reason, will react to naturally occurring conditions and in some instances such as the following, manmade conditions. Although rather bizarre this example is worthy of mentioning. The MFS, Jay Scott Ballinger, claims to have been guided by some satanic force. He may have operated under the guises of celestial and lunar events. The federal agent who interviewed Ballinger said that he would also set fires whenever he spotted the number "6" on a license, a road sign, and a billboard all at the same time. For these reasons and others, which were not explained, environmental conditions should be considered when plotting and analyzing calendars.

The environmental condition that most people are familiar with, the weather, will always have an effect on an individual's routine. Some MFSs will base their travel or activity outside on current or future weather conditions. These will all vary with the individual and the conditions. As mentioned earlier, seasons and weather are directly related and may have a significant effect on the success of the MFS. For example, during the early spring seasons in the northeastern US, the lack of green vegetation combines with the warm, dry, windy days and becomes the breeding grounds for novice and juvenile MFSs. These conditions, commonly known as a dry-spell, allow rapid fire growth in a target rich environment. This holds true for the rest of the US although dry spells occur during varying seasons depending on the region. Furthermore, consider that in most Northern regions of the US, the winter months tend to bring on freezing conditions with the possibility of snow. These cold, dark, and snowy nights are when the advanced and organized MFS will likely feel more confident. With visibility low and people cooped up in their residences trying to stay warm, the arsonist can move unnoticed among the few willing to brave the cold. Furthermore, the inclement weather will act to delay any eyewitness notification of the actual fire and delay fire department response. This will ensure the fire growth and the covering of the suspect's tracks. These are some of the obvious advantages of weather to the MFS. Do not automatically assume that they will be the only times for patterned fire activity. Some activity could occur when it is least expected. Consider the following case in point.

While researching MFS incidents, a case in upstate New York came to my attention. It involved two individuals who set fire to cars, sheds, and eventually occupied structures on days that it would rain. After performing some of the basic concepts of the process covered in this book (correlating, map plotting, analysis, and establishing patrols) the two subjects were apprehended. The motive for the fire setting activity was revenge against those who the two suspects felt had wronged them in their business transactions. Their MO was based on their occupations, that of landscaping. When it would rain they could not work. Therefore, those were the days that they would seek their revenge. Once again, use the concepts discussed in previous chapters for researching weather history. Weather information will be easy to plot and overlay. We will now explore how to use overlays or computer calendars to plot correlations between weather and fire activity.

Calendar plots discussed in this chapter are easy to design since the bulk of the work is usually already completed. The method used to develop a calendar plot is similar to the method used to develop clusters on a map and will take approximately 30 minutes to complete. It is time well spent. The recommended types of calendars to utilize in plotting are one-year, six-month, quarterly, or three-month calendars. Any variation of time periods can be used, but it is better not to use anything less than a three-month calendar plot. The reason for this is that the overall timeframes will be limited, making it difficult to identify the start dates verses breaks in activity. The months do not have to be within one calendar year, but they should be consecutive months. One-year calendars will obviously reveal the greatest potential for all incidents to be plotted and tracked. One-year calendars will help to eliminate confusion when a pattern is not immediately known, such as when street addresses are not always available or notifications are not always accurate (such as when received via wireless phone communication by individuals unfamiliar with the location). When it appears difficult to pick up a pattern, calendars can help by establishing clusters that occur on specific days or during weeks or months. These trends can help narrow the field or establish incidents to focus research on. One-year calendars will also help when you must search for previous incidents not known at the start of the investigation. The use of the one-year calendar will help in tracking by allowing for timeframe or cluster periods to become noticeable. The disadvantage to one-year and six-month calendar plots will be the amount of information displayed. Due to the need to display all 365 or 183 days, the space for additional information will be limited. (See Table 9-1).

For example, during the pattern search analysis of the map in Table 9-1, it can be noted that clusters appear to occur during the fourth week of every month. These incidents appear to occur over a three-day period. The majority of these three-day clusters occur on Thursday, Friday, and Saturday on mild, clear days. This trend appears over the six-month period of the search data. Knowing this, the six months prior to the primary search can be checked for incidents using the method discussed in chapter 7 and by factoring in the weather conditions over that period. This example shows how this technique could be used in addition to the other method as a standard by which to confirm the incident.

2003

January

S	M	T	W	T	F	S
			1	2	3	4
5	6	[7]	8	9	10	11
12	13	14	15	16	17	18
19	20	21	22	23	24	25
26	27	28	29	30	[31]	

February

S	M	T	W	T	F	S
						1
2	3	4	5	6	7	8
9	10	[11]	12	13	14	15
16	17	18	19	20	21	22
23	24	25	26	27	[28]	

March

S	M	T	W	T	F	S
						1
2	3	[4]	5	6	7	8
9	10	11	12	13	14	15
16	17	18	19	20	21	22
23	24	25	26	27	[28]	29
30	31					

April

S	M	T	W	T	F	S
		[1]	2	3	4	5
6	7	[8]	9	10	11	[12]
13	14	[15]	16	17	18	19
[20]	21	22	23	24	[25]	26
27	28	29	30			

May

S	M	T	W	T	F	S
				1	2	3
4	5	6	7	8	9	10
11	12	13	[14]	15	16	17
18	19	[20]	21	22	[23]	24
25	26	27	28	29	30	31

June

S	M	T	W	T	F	S
1	2	3	4	5	6	7
8	[9]	[10]	11	12	13	14
15	16	[17]	18	19	20	21
22	23	24	25	26	[27]	[28]
29	30					

July

S	M	T	W	T	F	S
		1	2	3	4	5
6	[7]	[8]	9	10	11	12
13	14	15	16	17	18	19
20	21	22	23	24	[25]	[26]
27	28	29	30	31		

August

S	M	T	W	T	F	S
					1	2
3	4	[5]	6	7	8	9
10	11	[12]	13	14	15	16
[17]	[18]	19	20	[21]	[22]	[23]
24	25	26	27	28	[29]	30
31						

September

S	M	T	W	T	F	S
	1	2	3	4	5	6
7	[8]	[9]	10	11	12	13
14	15	16	17	18	19	20
21	22	[23]	24	25	[26]	[27]
28	29	30				

October

S	M	T	W	T	F	S
			1	2	3	4
5	6	7	8	9	10	11
12	[13]	14	15	16	17	18
19	20	[21]	22	23	24	25
26	27	[28]	29	30	[31]	

November

S	M	T	W	T	F	S
						1
2	3	4	5	6	7	8
9	10	11	12	13	14	15
16	17	18	19	20	21	22
23	24	25	26	27	28	29
30						

December

S	M	T	W	T	F	S
	1	2	3	4	5	6
7	8	9	10	11	12	13
14	15	16	17	18	19	20
[21]	22	23	24	25	[26]	27
28	29	30	31			

Table 9–1:

Full-Year Calendar Plot. The dates that are circled identify days that incidents occurred.

Calendars of less than three or four months tend to confuse trends with patterns. This is especially true when a particular season is singled out. A trend related to weather conditions may be confused with a MFS activity. An example of this would be an increase in structure fires around the season when the onset of cold weather begins. This occurred in a primarily summer beach rental haven. During that particular rental season the fall weather arrived in early September. Because most of the residences located in this community were not heated, the renters began employing auxiliary heating units to stay warm at night. The combination of cold weather, laid back attitudes, and the consumption of alcohol lead to problems. When the number of fires increased in a short period of time within that community, the trend could have been miscalculated as a pattern. Once again, by combining the tracking process with the development of a calendar plot, the fire activity could have quickly been related to the weather, based on the early arrival of cold weather.

In keeping with the form of other chapters in this book, we will first discuss the handwritten method followed by the computer method of calendar plotting. Since you will usually make more than one calendar plot, it is recommended that you make additional copies of the calendar, either by hand using a pencil or by saving copies of the calendar in your computer (you can use the "Save As" option in your computer program and give each calendar a different name). The advantages and disadvantages of each method will also be discussed. Unlike the comparisons of handwritten and computerized mapping discussed in chapter 8, the comparison of the calendar plotting methods are not that clear. Both the handwritten method and the computer method have their advantages.

As mentioned in chapters 4 and 6, most sources of data will not identify the specific day of the week, only the numeric digit for any given month, day and year. Therefore, a calendar will be required to capture the specific day. If a full year calendar is used for this task, the date can be highlighted at the same time that the day is identified. When the incident is entered onto the calendar, a quick reference plot is created at the same time that the data is input. This is an advantage of a handwritten calendar plot over the computerized calendar plot because is the information can be plotted at the same it is entered into the database. This calendar plot may seem somewhat busy and should not be considered accurate due to the fact that the refining and weeding out of non-MFS incidents have not been fully completed. Do not be discouraged if nothing is identifiable based on this plot. Remember, this is based on preliminary data. Even if activity is readily noticeable in the plot, it is strongly recommended that reports are not based on this preliminary calendar for the reasons discussed in chapter 7. Having said that, the primary plot should help you to focus on when the activity began, what time period it peaked, and when the activity is most concentrated.

This first calendar can also help establish some preliminary observations such as when breaks in activity occurred and any cooling off periods. These breaks could last for days, weeks, or months. Many circumstances such as weather conditions or travel by the MFS out of the community will be factors in the plotting process. Overlays, colored pens, or highlighters can be used to show the weather conditions during the timeframe(s) in question. Once the data has been refined, you can do additional plots.

The first calendar plot will not be as easily completed with the computer version. Unless two computers (one with the database program and one running the calendar creator program are available) it will be difficult at the very least, if not impossible, to do both with one computer. However, this can be accomplished with one computer in a Windows-based operating system with high-speed processors or split screen technology. Unfortunately, most governmental agencies are not operating with this technology and their systems run extremely slow and make it nearly impossible to accomplish anything in a timely fashion. In this situation, the delay and waiting for the opening and closing of Windows will seem to only increase the man-hours required. In this case the handwritten calendar plot may save time. It may also be considered less threatening to someone who is not a common computer user. Either way, data entry is required.

If available, a full wall-size calendar is best, but could become cost-prohibitive when looking at more then one MFS case. One-year or six-month wall calendars can be purchased from any stationery store. Grease or erasable marker calendars can also be purchased, which can be found in all the monthly variations mentioned. The one-year version will allow for the greatest flexibility. For example, this calendar will allow for any period to be plotted up to a year and will allow for non-calendar years to be plotted. This would be most useful when a four-month period is placed in the center quadrant allowing for any period previous and thereafter to be plotted on either side of the original track. Therefore, if March to June were the focus of the research, the capability to plot back to November of the prior year and to October of the current year would be available. Unfortunately, these plots would be impossible to save for further analysis. The least expensive avenue would be to get a calendar that can be reproduced on a copy machine. These calendars can be found in the first few or last pages of personal date books. These simple year or six-month at-a-glance calendars can be copied and enlarged for additional plotting requirements. There are also computer software calendar creators available that are useful.

As noted in chapter 8, when plotting incidents on maps, whether the using the computer or doing the process by hand, there will be specific items that should be identified on the calendar plot. The first item will be the incident itself. How the fire activity is identified will vary with the method used. When conducting a handwritten plot on a one-page calendar, simply placing a circle around the numeric digit for that particular day in question will ensure easily identifiable marks to signify active days. When two incidents occur on a particular day, a slash can be placed through the circle to signify two incidents. If multiple incidents occur on the same particular day, then place another slash in the opposite direction of the first. This will create an "X" through the circle digit. These marks are easily recognized when conducting analysis for routines and patterns (see Table 9-2).

April							
S	M	T	W	T	F	S	
		1	2	3	4	5	The circles represent single incidents per day.
6	7	8	9	10	(11)	(12)	
13	14	15	16	17	18	19	The circles with one line represent two incidents per day.
(20)	21	22	23	24	(25)	26	
27	28	29	30	31			

August							
S	M	T	W	T	F	S	Circles with "X" represent multiple incidents per day.
					1	(2)	
(3)	4	5	6	7	(8)	9	Rectangles and squares represent rain days.
10	11	12	13	14	(15)	16	
(17)	18	19	20	21	22	23	
24	25	26	27	28	(29)	30	
31							

Table 9–2:

Calendar Marking Methods

When using a larger version of calendar (with space at each individual date to enter information) try to either circle the numeric digit or place a line under the digit. Use whatever form is most easily identifiable to the tracker. Although not required, it is recommended that the time of the incident be noted on that day by using different color pens or markers for each timeframe. For example, when using a one-year, one-page calendar, try identifying the morning hours in red, daytime or mid-day hours in blue, and nighttime hours in black. With a larger version of the calendar, try to continue the use of different colors by writing the information in military time (if possible) for each date. When deciding what timeframes to use, consider using timeframes that will be consistent with patrol schedules. For example, if the shifts are 3:00p.m. to 11:00p.m., then label the colors to coincide with that shift. This will allow for a breakdown by patrol areas. On the other hand, if the shifts run from 4:00p.m. to 12:00a.m., consider the following: 1:00a.m. to 9:00a.m. as morning, 9:00a.m. to 5:00p.m. as day-time, and 5:00p.m. to 1:00a.m. as evening. The reason for using these timeframes is so that a pattern will not be overlooked. For instance, if these timeframes are not used, an incident that occurs after 11:00p.m. on the first Friday and an incident that occurs after 12:00a.m. on the third Saturday will not show up as every other weekend night shift. Instead, they would show up as separate unre-lated time periods that could be missed by those conducting the analysis of the calendar and the end users will not draw and appro-priate conclusion in the final reports. The pattern would not be eas-ily identified or linked.

As with the computerized map plotting method, the com-puterized method of calendar plotting will continue to advance and become more user-friendly. Without dating this book before it is published, we will once again stick to the general require-ments of the software without getting into specific applications. The advantages of computerized calendar plotting are that inci-dent and information can be added to and deleted without hav-ing to re-enter or re-plot all the entries. This will become an important concern when dealing with a large quantity of inci-dents. Another advantage to computerized calendar plotting is the ability to produce additional calendars without the use of a photocopy machine. This can be very useful when employing a color printer for multicolor identifications.

There are many calendar creation software products available that can help, to various degrees, to accomplish the needs of the calendar plot. As with mapping software, calendar software varies widely in price and capabilities. Researching the software to find the one best suited to accomplish the plotting requirements may be dif-ficult. The first item noted would be that all the software researched will create calendars. What varies are the size, the style, the years available, and what can be added, deleted, and saved. When choosing calendar software programs, consider the ability of the software to enter data, details, or symbols in individual dates or days. Other requirements will be the ability to allow for the use of an overlay, to use of different colors (or shades), and to customize the calendar. Print size should also be considered, but remember that this feature will also be limited by the size of the printer. Computer calendar programs will vary between home and office use so be sure to research which type will best suit the plotting requirements. Many business- or office–designed software packages or bundles may contain some form of schedule planner or daily organizer that can be adapted to perform the calendar plotting. These programs will allow for data to be entered at specific days and/or times. The data entered is then displayed (and can be printed) in a month calendar form. If this feature is available, these individual months can be printed to create a full wall calendar. To help cut down on confusion, it would also be helpful to title each month with the type of data shown. This concept is similar to the maps and overlays discussed in chapter 8. If space is available, a full–wall calendar plot along with a large tracking map can be a powerful tool, not only as a status report board, but it may also be helpful to confront a MFS with a full–size calendar and map, showing their fire activity. This approach was used on Paul Keller by the investigators of the Arson Task Force in Washington State. The walls were plastered with fire scene photos, maps, and the artist's sketch of the suspect, along with other techniques that helped investigators gain a full confession from Paul Keller. If this method is being considered for interviewing, be sure to follow the principles discussed in chapter 8.

Once the technique for plotting has been decided, additional concepts can be considered. One of the best concepts for calendar plots is to use overlays. As mentioned earlier, significant events (such as weather conditions and altered routines) can be developed as copies or overlays of the original plot. These help weed out and refine the plot. For example, items such as lightning strikes, storms packing high winds, or extremely dry weather conditions can be plotted to help rule out multiple event days. Other types of overlays or copies would include (but are not limited to) work schedules, school schedules, weather conditions, association meetings, holidays, as well as political, religious, and extremist activities. The placement of this refining overlay on top of the incidents plotted (as part of this process) will only be performed once, yet it will be employed in the analysis of all suspects.

These overlays or copies are extremely useful when more than just one unrelated suspect is being investigated or when subjects may be working together to set fires. For plotting multiple subjects, continue to use separate overlays for each individual (See Table 9-3). If the potential for the unsub to act in concert with others exists, be certain that multiple overlays will be legible when placed on top of one another. Use the same types of methods discussed for creating the handwritten copies in chapter 8.

The items needed for handwritten overlays will include the refined calendar plots and clear plastic transparencies that completely cover the calendar or month, as well as pens or markers that can write on the transparencies and a photocopy machine. Start by marking the top and/or bottom of the overlay to coincide with the calendar. Place the overlay on top of the refined calendar plot and identify an open area on the overlay for the title. This title should be visible when viewed on top of the refined plot or placed between other overlays. Look toward the top and side margins. Once completed, use the overlay to identify information pertinent to the suspect and the incidents. Use pens and/or markers that will stand out from the refined plot. Caution should once again be mentioned. Do not limit or omit information from the suspect plot because it does not fit the plot. Do not try to fit the suspect to a crime they did not commit. If necessary, use an unmarked calendar first with the overlay to indicate all the information related to the suspect without the influence of the data plotted. Once all related data is plotted, place the overlay on the calendar plot of the fire incidents. The suspect information and fire data can than be cross-referenced without bias.

2003

January

S	M	T	W	T	F	S
			1	2	3	4
5	6	[7]	8	9	10	11
12	13	14	15	16	17	18
[19	20	21]	22	23	24	25
26	27	28	29	30	31	

February

S	M	T	W	T	F	S
						1
2	3	4	5	6	7	8
9	10	[11]	12	13	14	15
16	17	18	19	20	21	22
23	24	25	26	27	[28]	

March

S	M	T	W	T	F	S
						1
2	3	4	5	6	7	8
9	[10	11]	12	13	14	15
16	17	18	19	20	21	22
23	24	25	[26	27	28]	29
30	31					

April

S	M	T	W	T	F	S
		[1]	2	3	4	5
6	7	[8]	9	10	11	[12]
13	14	[15]	16	17	18	19
[20]	[21]	[22]	23	24	[25]	26
27	28	29	30	31		

May

S	M	T	W	T	F	S
				1	2	3
4	5	6	7	8	9	10
11	12	13	[14]	15	16	17
18	19	[20]	21	22	[23]	24
25	26	27	28	29	30	31

June

S	M	T	W	T	F	S
1	2	3	4	5	6	7
8	[9]	[10]	11	12	13	14
15	16	[17]	18	19	20	21
22	23	24	25	26	[27]	[28]
29	30					

July

S	M	T	W	T	F	S
		1	2	3	4	5
6	[7]	[8]	9	10	11	12
13	14	15	16	17	18	19
20	21	[22	23]	24	[25]	[26]
27	28	29	30	31		

August

S	M	T	W	T	F	S
					1	2
[3]	[4]	5	6	7	8	9
10	11	[12]	13	14	15	16
[17]	[18]	19	20	[21]	[22]	[23]
24	25	26	27	28	[29]	30
31						

September

S	M	T	W	T	F	S
	1	2	3	4	5	6
7	[8]	[9]	10	11	12	13
14	15	16	17	18	19	20
21	22	[23]	24	25	[26]	[27]
28	29	30				

October

S	M	T	W	T	F	S
			1	2	3	4
5	6	7	8	9	10	11
12	[13]	14	15	16	17	18
19	20	[21]	22	23	24	25
26	27	[28]	29	30	[31]	

November

S	M	T	W	T	F	S
						1
2	3	4	5	6	7	8
9	10	11	12	13	14	15
16	17	18	19	20	21	22
23	24	25	26	27	28	29
30						

December

S	M	T	W	T	F	S
	1	2	3	4	5	6
7	8	9	10	11	12	13
14	15	16	17	18	19	20
[21]	22	23	24	25	[26]	27
28	29	30	31			

Table 9–3:

Full-year Calendar Depicting Probable MFS Incidents. Possible reasons for lack of activity on certain dates are shown by the rectangles and boxed dates that represent rain days and the hunting season.

For computerized overlays, the ability to save the basic (or primary) data will be required. This basic data will be the template for all other overlays. The basic data should include the incident types, their time, and if possible, abbreviations for location. Once again, when considering software, see if this feature is available. If not, it may be necessary to combine the methods, such as creating and printing the basic data and then creating the overlays by hand. The important thing is creating a good quality, usable end–product. When employing a purely computerized method, continue to abide by the information for the handwritten method. For example, continue to save the primary data in the original file; then layer the additional data over the primary. At each layer, save the additional data as a separate file, so that when plotting three separate issues there will be four separately saved files. The first file will be the primary data; the second file will be the primary plot with the environmental overlay (forming the refined plot); the third file will be the refined plot; finally, the fourth file will be the refined plot with the suspect's data plotted.

To assist in plotting the computerized version, try to develop a symbol or abbreviation for some of the required data. For example, some calendar software will contain canned or pre-drawn symbols. Use these symbols to designate certain types of fires and/or environmental conditions. If symbols are not available or occupy too much space, try using abbreviations for the types and conditions. For example, grass or brush fires could be abbreviated to "BR", Vehicle fires to "VE" and so on. Phases of the moon could be abbreviated to "FM" for full moon and "NM" for new moon. This would also be useful for extreme temperatures and/or precipitation. Some calendar software will automatically place needed information onto the finished product. This may include phases of the moon, recognized national holidays and significant almanac information. With the completion of the calendar plot and the data analysis, the potential of MFS activity can be established, conclusions will be drawn and the decisions for further action can be determined.

Chapter 10

WHAT TO DO WITH THE RESULTS

Now that the information has been collected, analyzed, plotted, and formulated into results, what should be done with it all? One of the greatest problems in public safety (and probably most institutions) is the failure to communicate in order to get the proper information to the individuals who need to know it so that they can do the job. This is the whole reason for this process. If no effort is going to be made to disseminate the information to those who do the job, then don't bother finishing this book.

While writing this book, I had a conversation with a fire investigator in upstate New York who had just led a task force that had arrested some MFSs. During our conversation, the investigator made the unsolicited statement that they had come to the following conclusion: the best chances for an arrest of the unsub(s) would be with a patrol officer who would be responding to or patrolling the area of the next fire incident. Once all the agencies involved (approximately six) realized this, the task force made sure that all the patrol units and the special patrol units in the active fire area were notified. This is true of most serial crimes. Competent, professional individuals who have been alerted to a situation will take note of unusual activity and will catch the bad guys. A few cases in point:

- In 1996, Joel Rifkin, the serial killer, was caught by a NY State Trooper during a car stop in the early morning hours on a road leading away from a densely wooded area.
- In 1977 David Berkowitz, the serial killer, was investigated in Yonkers, NY, for a parking ticket, which placed him in Brooklyn, NY, late at night near a double shooting. This was nowhere near his home.
- In 1999, in the state of Indiana, paramedics became suspicious of Jay Scott Ballinger when he waited two days before seeking treatment for burns he received. On a tip from the paramedics, a patrol officer responded to the hospital, interviewed Ballinger about the burns, at which time he admitted to setting a fire. It was later learned that he had set over 50 fires, including one that caused the death of a firefighter. Ballinger claimed to worship Satan.

Remember to keep the concept of tracking within the context of this book. Do not confuse the tracking process with basic police work. The process will help to identify, develop, and to some extent make the case, but after alerting all appropriate personnel, know that someone else will make the arrest.

For some investigators, the thought of sharing information is one of the ultimate sins. This is not a valid concept for MFS investigations. When dealing with MFSs, it will be important to share information with others; therefore, think outside the box. Keep all options and communications open with personnel and agencies that would not usually participate in the standard investigation. This is not to say that you should give the store away, but that certain items pertaining to the case will need to be shared. Most of the details that need to be shared are public information anyway. The tracking report will contain generalizations about the overall case. Specific items as well as sensitive details can be included but are not usually required. When there are multiple agencies and/or personnel receiving the report with varying need to know, some changes may be required. Therefore, modified versions of the report can also be disseminated.

Examples of resources would include probation and truancy officers, as well as administrators. Probation officers can help to place a name with fire activity in a particular area, thus helping to identify persons convicted of similar crimes and narrowing the search to known subjects who live in or around the active fire area. In many cases, probation officers will have insight to a subject's schedule and social habits, as well as who may have been recently released. From this it is possible to identify pattern searches and calendar plots that match up with their probation subject. Truancy officers may know the locations where different youth gangs and groups frequent and also identify who the chronic absentees are and who will offer the greatest information.

Additionally, do not neglect to remember that when disseminating information out, it should also go up. Sending information up the chain of command to administrators will play a key role when requests involve assistance from personnel outside the standard command or from outside the agencies. This will help to keep the bosses ahead of the curve. A well-informed administrator is usually a happy administrator. Disseminating information to administrators will also help with funding for future tracking cases. Also, if a "profiler" is employed, this information will be valuable to him or her. Be sure to keep them in the information loop and be open to other numerous contacts.

By way of an example, a report could identify that a flare was used during the fire setting. This would be helpful to any patrol officers who interviewed anyone or conducted traffic stops around the active fire area. No specifics will be mentioned on how the flares are used to set the fire, where they were found, or what types of flares were used. This will protect any information related to the suspect's signature. The same can be said for accelerant products, be they small containers of flammable liquid, combustible glues, or shredded paper. The "how", "where", and "what" are retained for further investigation and interviewing needs. The inner circle of shared information should not just be limited to fellow investigators or patrol officers.

Fire suppression personnel should also be considered. The reason for this is simple. The fire department receives the call; if they do not know to keep the investigators informed, chances are good that they will not be. For instance, in all the cases that we cross-referenced with the police blotter or data bank at least 20 to 25 percent of the fire incidents were not noted. This could be due to a number of reasons. The officer was on another call and not available. Covering units were not notified. Calls did not initiate through the 911 system. Relief officers did not know to notify investigators. For whatever the reason, the fact remains the same. Not all the incidents make it to the system and therefore not all the incidents make it to the investigator.

This does not mean that the entire fire department needs to know. In fact, it should be limited to a select few. For example, in a paid department, certain personnel will probably work on shifts that are identified as active. Every shift will have a supervisor. This supervisor could be notified to inform investigators of certain activity at specific times in certain areas of the district. In a volunteer department, this would be the chief officers or designated incident commanders. This is usually a subordinate officer. Their assignment could be to notify the chief, who will notify the investigator. For some investigators, this will be the ultimate leap of fate. The prevailing thought is that this will be the fastest way to screw up a case and there may be some validity to that concern in some cases. After all, loose lips sink ships, but sharing the basics may help to reveal some basic facts. Consider the following observations:

1. If the information does leak, then obviously the suppression personnel cannot be trusted. Depending on your approach to the situation, it also means that this could be a source to disseminate false information.
2. If the fires stop, then two primary goals of this process will ultimately have been accomplished; a MFS will have been identified and thwarted. With no further incidents, the investigation will continue without the added pressure of further incidents. Some may feel that this will hinder the investigation, but it should be seized upon to turn up the pressure because obviously the unsub is scared and fears being caught.
3. Even if the fires continue, but in a new location, then the word or data is getting out and the circle of suspects has quickly narrowed to someone active within the fire service community or relatives of fire personnel.

4. The information may not leak, intelligence may be gained, and fire service personnel may turn into good HI sources for future reference. My personal experience is that this is how it usually works out. The use of these ideas will be expanded upon in chapter 12, but for now realize that the sharing of basic general public information is not always a bad thing.

Having said all that, you nevertheless should remind personnel that access to your findings is somewhat limited, restricted, or even classified. Although classified is a strong term and would usually involve further checks and security procedures (such as numbering and signing for each copy), it would be somewhat difficult to enforce in this process. Nonetheless, a reminder within the document expressing the concern is a good mental re-enforcement. This will act to advise the reader that not all personnel or persons need to see the document, nor does it need to be discussed with everyone. It is a fine line between who needs to know and who does not, but it is not an unrecognizable line. It is important to spell out up front who will receive this report so that the players will know who they can discuss the report with and who they cannot.

So now that we have settled why we give out the information, the next issue to address is how to disseminate. The first step is to develop a report that is simple to understand, legible, and easy to reproduce. It should be as few pages as possible (usually one to three pages) typed on standard $8^1/_2$ by 11-inch paper. Keep in mind that this report will be read by fellow investigators, trackers, analysts, supervisors, administrators, and possibly outside agencies; therefore, have the report proofread and check for proper grammar and spelling of addresses and titles. This report is not a commentary or the editorial page of the local paper. It will be a dry, to the point, unbiased, and objective report. Strictly use the data and do not ad lib. Let the facts tell the story. If there is no activity in the month of May, say so. When activity is noted in a particular area in a particular month, simply state it and do not try to explain it; just stick to the facts. The report should start with a title or name, such as Fire Incident Tracking System or FITS. This should be followed by the following basic information: the date of the report, who produced it, the time period that this report covers, the area(s) in question, who should receive the report, and finally, the point of contact or clearinghouse where the reader can report further information (See Table 10-1).

FITS REPORT
FIRE INCIDENT TRACKING SYSTEM REPORT

PAGE 1 OF PAGES 1

REPORT ORIGINATED BY B. Martinez FRES	REPORT # 001	COMPLAINT NO. 00-010101	DATE OF THIS REPORT 2/15/00 to 6/7/00
TOWNSHIP Islip	SCHOOL DISTRICT Central Region	HAMLET Littleville	FIRE DEPT. Littleville	TYPE OF INCIDENTS auto, debris/ brush, dump, struct

PRECINCT SECTORS 3rd 307 & 308	DATE OF NOTIFICATION 5/31/00	AGENCIES NOTIFIED County Arson Task Force, County Police, State Police

The analysis of some 50 Incidents occurring in the Littleville Fire District, from 2/15/00 through 6/1/00 have revealed the following patterns.

NOTE: The information that follows is not a profile but a study of the facts surrounding these incidents. These facts include the time, date, day of week, and location of the incidents. The information in this report has been gathered from two independent data sources (Fire Rescue Communications Log and the Littleville Fire Dept. Incident Report Database).

FACTS:
Fires in the Littleville Fire District are concentrated to three clusters. The clusters in question have been broken up into three separate areas. Refer to the attached map for specific details. The areas have been identified as follows.

1.   Main street (Rt. 25) near Maple St. Grover La. and 5th Ave.
2.   Walter's Elementary School Area
3.   Freeman Hwy (Rt. 347) around Lake St. (to the North) and Rustic Rd (to the south)

Of the 54 incidents the majority occurred in the Littleville Fire District. Fires in the neighboring Fire Districts of Middleton and Burger do not show any patterns or concentration similar to Littleville. The incidents researched involve autos, brush, debris, dumpster, and structure fires. All structure fires in question have occurred after dumpster fires in the same area approximately 2 to 3 hours before the structure fires in the Main St. area. Highest frequency of incidents occur on weeks of winter and spring break of the Central Region School District. No fire activity was noted on days with precipitation except for activity near Lake St. and Rustic Rd (in the Freeman Hwy. area). In all three cases the activity occurred on Saturdays during the early evening hours of the precipitation days.

As for specific days the analysis revealed the busiest days appear to be Friday, Saturday, and Sunday with brush fires being the prime target in the Main Street area. This includes multiple incidents each day. As for the timeframe, brush fire activity started from 2300 to 0300. At the Walter's Elementary School area, the activity also occurs on weekends in the afternoon to the evening hours (around 1830). with brush and cars being the primary target. An MFA call occurred at 0300 on 11/14/99 at a pay phone along Main St. this was consistent with other incidents that occurred in the early morning hours along the Main Street area during 1999 and 2000. The caller was said to sound drunk.

Structure fires had the smallest time frame window for occurrence. Starting after 1945 to 0030. Only three structure fires were noted.

Table 10–1:
Example of a Fire Tracking Report

The detailed information of the report should include the following items in the following recommended order:

- The disclaimer to explain what the report is
- What the report is based on
- Most importantly, what the report is not

In particular, that it is not a profile, nor is it a field or case report. Then explain what it is intended for (similar to a mission statement) and why this report is being generated. For example, due to the unusually high number of fire incidents or the increased level of violence noted in the fire activity. Furthermore, state whether the goal of this report is to stop further incidents, to assist the investigation, or accomplish both missions. The best way to state this is to explain that this report is based solely on the facts that surround these incidents and have focused on the times, dates, and days of the week, locations, and types of incidents. Next an explanation of the meta data as discussed in chapter 4 should be noted. This explanation relates back to the brief explanation on where, what this data is based on, and the number of incidents researched (See Table 10-2).

The analysis of some 50 Incidents occurring in the Littleville Fire District, from 2/15/00 through 6/1/00 have revealed the following patterns.

NOTE: The information that follows is not a profile but a study of the facts surrounding these incidents. These facts include the time, date, day of week and location of the incidents. This information has been gathered for investigative purposes and should be limited to the agencies noted above. The information in this report has been gathered from two independent data sources (Fire, Rescue Communications Log and the Littleville Fire Dept. Incident Report Database).

Table 10–2: **Details of the tracking reports disclaimer, the purpose of the report, and the source of the data.**

Next, report how many incidents the track was narrowed down to as possible, then further discriminated to as probable and known MFS incidents. This is important for a number of reasons. The first reason is because for many parties this will be the first time that they have received a report such as this. To remove as much doubt as possible, let them know where these conclusions are coming from. As Kaiser would say, "Cops are grounded guys that do not believe in the boogy man." If you think that the guy killed his wife, he probably did, but you had better prove it and not go off on some wild goose chase based on as of yet unproven information. Secondly, those who receive the report can also research the information for themselves and see if what this report says is true. Sometimes people need to be reminded that this is not wizardry or rocket science. If there is any doubt, someone will work to disprove it. The third reason is so that the reader of this report will not assume that the databases or resources used are those that he or she is familiar with. Hopefully the individuals reading the report will inform the designated point of contact of any additional information or intelligence sources that can be considered.

The next part of the report is to identify the active fire areas by the locations of the incidents. Establish the activity areas with names familiar to the area or community. Use terms or titles for neighborhoods, developments, and industrial areas that are well known to the responders. If there are no names for the active fire area, use a street name that runs through the designated areas. This would be the main connecting road or main drag, which could also be considered the popular hang out. If there are multiple cluster zones within the active fire area, try to also identify them by area names and not by compass points or sectors. This may be important based on the fact that public safety personnel (such as patrol officers or EMS units) who work within the area may know something about these city blocks or neighborhoods. If the areas are known by a particular name such as the Meadows or the Heights, you will have to explain where these are located.

In all cases, reinforce the need to conduct searches at the local level, or at the very least, encourage local participation. The more personnel who have input into the development of the report, the greater will be the synergistic effect to discover all the facts. If necessary, use the patrol sector designation as a last resort. When a particular name for a designated area is considered not politically correct, it is recommended to rethink the use of that particular name for a report. Whatever name is chosen, try to use one name that cannot be confused with another. For example, some areas may encompass gang territory or "turf". Gangs have been known to use fire as a weapon during turf wars, burning property and vehicles of rival gangs along turf boarders (what is considered the turf of one gang may also be disputed and considered the turf of another). The goal is to intimidate people of the area to join their gang and force the rival gang out. In one case, rival gang members would enter another turf and set fire to the opposing gang's (or that member's family) vehicle. The intent was to get the rival gang members out into the street to attempt a drive-by shooting.

The next step is to discuss the types of incidents that have occurred. Once again, breakdown to specifics such as the starting type or original type of incidents confronted. Designate the time-frame for these types of incidents. Identify if certain incidents cover specific periods and any dormant periods for these incidents. Also establish if there is a definitive escalation of violence with the types of incidents or if the activity has remained constant with only one type of fire occurring consistently. This section of the report information, along with location, will help the memories of patrol, suppression, and EMS personnel to recall previous activity that may be pertinent to the case.

The next issue to address is the times of the occurrences and the days that they occurred. Breakdown the most and least popular times and days. If there are particular types of incidents occurring at specific timeframes and/or on a certain day, address these items as well. Next, discuss the monthly activity trends. Be sure to cover any specific periods of inactivity or high activity. Report any trends related to the busiest weeks, if the unsub prefers the beginning of the month, the middle, or end. If the third week is consistently busy, note it. Note any holidays or almanac events that may be related to a particular day's incidents.

Finally, identify any environmental factors that may have played any role during the noted periods. Try to relate everything back to the types of incidents, times, and locations by referencing the map. Relate all these issues to the titles or names designated for each particular area (See Table 10-3).

FACTS:

Fires in the Littleville Fire District are concentrated to three clusters. The clusters in question have been broken up into three separate areas. Refer to the attached map for specific details. The areas have been identified as follows.

1. Main street (Rt. 25) near Maple St., Grover La., and 5th Ave.
2. Walter's Elementary School Area
3. Freeman Hwy (Rt. 347) around Lake St. (to the North) and Rustic Rd (to the south).

Of the 54 incidents the majority occurred in the Littleville Fire District. Fires in the neighboring Fire Districts of Middleton and Burger do not show any patterns or concentration similar to Littleville. The incidents researched involve autos, brush, debris, dumpster, and structure fires. All structure fires in question have occurred after dumpster fires in the same area approximately 2 to 3 hours before the structure fires in the Main St. area. Highest frequency of incident occurs on weeks of winter and spring break of the Central Region School District. No fire activity was noted on days with precipitation except for activity near Lake St. and Rustic Rd. (in the Freeman Hwy. area). In all three cases the activity occurred on Saturdays during the early evening hours of the precipitation days.

As for specific days the analysis revealed the busiest days appear to be Friday, Saturdays, and Sunday with brush fires being the prime target in the Main Street area. This includes multiple incidents each day. As for the timeframe, brush fire activity started from 2300 to 0300. At the Walter's Elementary School area, the activity also occurs on weekends in the afternoon to the evening hours (around 1830). with brush and cars being the primary target. An MFA call occurred at 0300 on 11/14/99 at a pay phone along Main St. this was consistent with other incidents that occurred in the early morning hours along the Main Street area during 1999 and 2000. The caller was said to sound drunk.

Structure fires had the smallest timeframe window for occurrence. Starting after 1945 to 0030. Only three structure fires were noted.

Table 10–3:

Incident Location Details

The next section of the report to discuss is the conclusions drawn. These conclusions are based on the total pattern analysis and overall tracks. For this report, the conclusions are an abbreviation or summary of the facts covered earlier. The conclusions should be written so that they can be used as a quick reference source or field guide for patrol and investigative personnel to fire activity. If possible, establish the conclusions based on the days of the week. This will allow the reader to not only pay attention to areas during their work hours; it will also allow for arranged schedules to concentrate on that particular area for clues, tips, and the daily activity noted. The recommended arrangement is by day, followed by the hours of activity with locations (or names of the active fire area), then the types of incidents to expect, and finally, where that particular day rates among the other six. Use designations such as: most active day, least active day, or potential for multiple incidents that day during the timeframe established (See Table 10-4). If a particular week or period within the month is consistently active, it should also be noted in the conclusions.

CONCLUSIONS:

Mondays - Look for incidents occurring around the Main St. area during the evening hours after 1830 to 2330. Look for auto and dumpster activity.

Tuesdays - Look for incidents to occur during the hours of 1630 to 1800 except on 4/13/99 when incidents started at 1147. Look for activity near the Walter's Elementary School Area with brush fires being the most frequent.

Wednesday - Look for incidents occurring in afternoon hours near the Walter's Elementary School Area with brush being the main target. This is the Least active of day.

Thursdays - Look for incidents to occur from 1200 and end by 1900 near the Main Street area.

Fridays - Look for incidents to occur from 1600 and ended by 0400 in the Freeman Hwy. area and the Main Street Area. Look or multiple incidents.

Saturdays - This is the busiest day with incidents starting at 1200 to 1830 in the Walter's Elementary School Area and near Main St. from 2000 to 0300. Most incidents occurring after 2300.

Sunday -Incident start around 0000 to 0300 to occur near Main St. with dumpsters the main target and from 1200 to 1830 to occur in the Walter's Elementary School Area with brush fires the main target.

Table 10–4:
Report Conclusions

The next item to consider for the report will be the map(s). Choosing the proper map design for the report can be challenging. One of the first things to consider is what the map needs to address. Depending on who receives the report will determine what type of map to include. Determine whether the map will be qualitative or quantitative. A qualitative map will show non-numerical data such as the types of incidents, the times, days, and boundaries of active fire areas. It will highlight areas of concern, clusters, and if possible, identify ground zero. The map will show details for access to and from, possible observation points, and can be used to determine the proximity of activity to probable suspects. This map will be most useful to investigators and patrol personnel, who will be able to employ it in interviewing and establishing patrol of the target area as well as further investigation of the scenes. If available, aerial photos can be used to supplement the qualitative map (See Map 12-2).

After identifying what the map(s) will address, the next step is to decide what type of map should be printed. This may help solve many of the issues surrounding the qualitative verses quantitative. Quantitative maps will portray the numerical information or the total number of incidents. This map can be somewhat busy and difficult to use to for the distinguishing of details. This is not always a disadvantage based on the idea of who will be viewing this map. For example, it may be recommended to display this map to the administrators who will be able to view the totality of the situation. Their requirements will be well suited to a quantitative map, which would display the big picture to them and not the specifics to the track. Quantitative maps may also be more useful to other analysts, trackers, and profilers, who prefer to conduct further research, analysis and other conclusions. This is also the recommended style to display to the media and in turn the general public if the investigation escalates or spirals into those directions. The reason for this is so that they will be able to grasp the totality of the situation, but will not be able to identify anything else other then a general idea of the locations. Keeping the bulk of the information to the investigation (See Map 8-3).

As far as fire suppression or prevention personnel are concerned, both qualitative and quantitative maps will be useful. For suppression officers, the qualitative map will be useful for assigning assets and future logistic needs. The quantitative map will allow the fire prevention officer to focus on factors affecting notification, propagation and probable hazards to suppression operations. If varying versions of the report are going to be produced, consider also producing varying map designs.

Next, decide what scale is best for your audience. For example, if you zoom out or increase the scale to say 1 inch equals 1000 feet, the detail will be decreased, but the bigger picture will be displayed, allowing the viewers to identify the community where the incidents occurred. Zooming in to where 1 inch equals 100 feet will allow for greater detail of all of the items discussed under the qualitative view. Whether one map or two are to be used, be sure that the map(s) will illustrate the issues discussed in the report, in particular, the majority of the issues discussed in the conclusions. If Thursday nights are the most active, try to identify the incidents and their locations. When there are two or more clusters, try to display all of them with some reference to their size and distances between each.

Once the type, scale, and design are chosen, the next step is to add the basic items required needed to read a map. This would include the title, legend, scale and orientation. The title should be related to the community, active fire area, or name of the track. The legend should include any road designations and geographic or topographic symbols used. Scale and orientation are simple, but just as important as any other requirement. Scale will allow the reader to further analyze the map for specific concerns related to their responsibilities. Orientation will make the map user friendly to all. Usually a simple arrow with the letter "N" pointing north will suffice.

The final decision will be whether the maps are produced in color or gray scale. In most cases, a gray scale map will display all the pertinent information just as well as a color map will, at a cost usually considerably less to produce and reproduce. When deciding whether to use gray scale or color ask yourself the following question: "Which can be most easily produced and will still best distinguish the issues covered in the written report? When the report distribution is being purposely limited (which should be in the majority of cases), which map type will be most difficult to reproduce?"

Finally, it is necessary to discuss how to disseminate and update the report. Once again, use the concept of advising on a need to know basis. Disseminate the report at least once, preferably in person. The reason for this is to confirm that the intended audience receives it and any questions related to the track can be answered immediately or relayed back to the tracker. When multiple shifts make it difficult to disseminate information, try to brief the supervisor and leave the reports with them to disseminate. As for facsimile, electronic mail, or postal service, try to avoid these forms of communication unless they can be considered secure. To help keep units that work varying shifts informed, establish a bulletin board (in a secure area) that can be updated daily or weekly. If the bulletin board is inappropriate for that setting, try utilizing a loose leafloose-leaf notebook to store the active tracks. Place this notebook at the occupied main desk or an area accessible to only those personnel who are immediately involved in the case. Try to avoid file folders within file cabinets, as they tend to be historical and not current or active. Be sure to leave a point of contact (with phone number) on the report so information can flow in both directions.

Chapter 11

WHAT TYPE OF EQUIPMENT IS REQUIRED TO TRACK MULTIPLE FIRE SETTERS?

With the rapid developments in electronics and the advances in the computer software industry, the use of new technologies to help fight crime has become a reality. We could spend chapters discussing what new technology will do for us. For two reasons we will not do that in this book. The first reason is because it is not the intent of the author of this book for the information in it to become outdated before it is even published. It is inevitable that new technology will be produced that will surpass what we currently consider state-of-the-art. The second reason is that although the use of technology such as Geographic Information Systems (GIS) and digitized aerial photography are becoming more and more common in local government, the dissemination of this information has been slow. This is especially important to most law enforcement street cops because in most cases the personnel on the front lines are the last to see any new technology, and by that time it can be five to ten years since it was introduced. In other words, old technology is considered "new technology" and truly state-of-the-art is beyond the reach of public servants for many years.

For these reasons, cutting edge technology and specific hardware will not be discussed. Instead, one of the goals of this chapter will be to discuss what is needed based on readily available, standardized equipment and how to utilize it. With this in mind, the technology covered will be limited to "off the shelf" products and even that technology will be covered to a limited extent. Cost for this equipment will also be discussed in general terms. Cost should not be simply thought of as monetary expense alone. There will be many other costs that this chapter will try to address. If the capital required to obtain computers is not available, then the handwritten method will be the alternative option. As always this chapter will begin by explaining the handwritten method. Because this method will be considerably less expensive in equipment, the cost needs will be covered as a time factor. Large quantities of time will be sacrificed in lieu of high costing equipment. Man-hours will be considerable and the chance for errors in analysis will increase. For example, without a computer for analysis, plotting a case involving thirty incidents using the handwritten method will take approximately two days. That means, at the very least, 16 man-hours to correlate and analyze for a seasoned tracker. Note the time to collect this data is not included in this figure.

Depending on whether salaried employees or volunteers are used, the expense in time and money will be considerable. Based on this fact, the additional cost will be the fatigue factor to the analysis personnel if they are also the data entry personnel. Remember that for many employers, time is money and longer man-hours will mean more fatigue leading to more errors and more time. As many fine lecturers and professors have said, the human mind can only absorb as much as human buttocks can endure.

Before we continue the discussion of each method, there are some general requirements for the office and its supplies. One of the biggest requirements will be workspace, preferably separate from the normal work area. This space does not have to be in a completely separate building or office, but it should be in a relatively secure area where items such as maps, photos, and reports can be displayed without fear of compromising the overall case. In the perfect world discussed earlier in this book, this space would be similar to the "Bat Cave" of the comic book super hero Batman.

Something a little more reality-based might be the US Marshal Services' mobile field command (nicknamed "Red October"). This unit has been used during mass fatality incidents, as well as field operations related to fugitive tracking. But to be totally realistic you can hope that your new tracker office will not be a desk, in the basement, under a leaky pipe, next to the heating system. Hopefully, some happy median could be acquired.

A good standard for room size would be the following. A minimum room size of approximately 15' x 20' to allow ample wall space for at least two workstations, a tabletop, and bulletin boards. Windows are not necessary, so converted spaces may be an option. Keep in mind the idea of security when developing this room and the posting of information.

Consider the amount of electronic equipment that will be in the room. When dealing with electronic equipment such as fax machines, copiers, TV sets, computers, radios, and scanners in any space three problems will rapidly come to light. The first is power sources, check to see if the space will support the power necessary. The second is the heat that all these systems will generate. See that adequate ventilation exists to allow for proper system operation and staff comfort. The third will be the radio frequency interference created by all these units in close proximity to each other. Some type of shielding may be necessary. All these issues should be addressed before arranging and setting up the room space. If additional computer equipment above the normal workstation requirements, such as wide carriage plotting printers or copiers were utilized in this space, then the dimensions would increase. Unfortunately, for some agencies space is a commodity that comes at premium cost. Do not become disillusioned and if necessary, make due with what is available, even if it is the desk under the leaky pipe.

Once the process is established and proven, it will be easy to justify better workspace. Then it can truly be said that this process started from the ground up. Once the space is acquired, it will need to be filled with the appropriate items such as a worktable and chairs. The table should be large enough to lay out maps and photos and analyze data. Try to avoid using a functioning desk or computer workstation. Although the computer station will be necessary (with the computerized version) it should not be considered part of the table space. These spaces will become insufficient quickly and will interfere with other ongoing work. A 4' x 8' bulletin board is also needed. If you do not have enough room for one this size, a portable style may be used. The bulletin board should be large enough for the track information or multiple tracks to be posted.

You will also need telephone lines, preferably with voice mail or an answering service (human or otherwise). Existing lines can be utilized if not currently overtaxed. Access to a fax machine should also be considered. If a computer is used with access to the Internet, consider a dedicated line for the system. Other communication equipment is also a necessity, especially in the form of secure wireless systems. In some cases, the MFS will use radio frequency scanners to listen in on fire ground and police communications. Secure channels or digital wireless telephone communication equipment would help avoid any chance of tipping the investigation. Wireless secure phones will also allow the investigator to follow through on other work-related matters while conducting surveillance or interviews. Sooner or later technology will be developed to make these systems less secure; therefore, communications over any wireless system should be limited with hard line systems being the recommended choice for detailed conversation.

Finally, the general office equipment such as desks, lights, chairs, paper, pens, and so on will be required. The last item you need for either method is the copy machine. When conducting this process by hand, this machine will be the greatest time saver. As for the computerized method, a copier will also help save the printer and increase reproduction time. The copy machine must be capable of enlarging, shrinking, and copying photographs without distortion. Most copy machines will now enlarge and reduce with little loss to quality. Some also sort and staple. The last two features would not be required, but should be considered. Color copiers are a tremendous step forward, but will also come at considerable expense, machine size, and production time. Whether a color copier or black and white machine is chosen, the issues of maintenance and supplies must be addressed in order to continue the productivity. Maintenance with color machines and the type of paper used could be expensive. Check to see which companies have the best lease, purchase, and service contracts. Depending on the method used, this item will be one of the largest expenses to obtain and maintain. If access to a capable machine exists, use it. Once again, speak with the agency or resident expert on copy machines.

An additional consideration would be clerical help. A clerical staff will help move the process along especially if it is being conducted without the aid of computers. It is always helpful to have someone to conduct the data entry, sorting, filing, and collating while the trackers care for the business of collecting and analyzing data. It would also help to have staff available to take messages. As stated in the beginning of this book, the information required has and will continue to be gathered but may not be attainable due to the limited time of personnel and access to record storage. When a staff member is available to follow up on research records, it will free up the time needed to gather other forms of intelligence.

Unfortunately, good clerical staffs are extremely cost prohibitive and hard to find; therefore, keep the following ideas in mind. Intern programs at local colleges or universities can be a very good source of interns who can care for clerical needs. These institutions may already have funded intern programs that will pay the students or give them credit toward their degree. It would be an even greater asset if these students were in degree programs that are somewhat related to the industry. Such degree programs would include economics, geography, crime mapping, fire science, and forensic science programs. Viable candidates for internships could also major in fields such as behavioral science (mental health), criminal justice, psychology, or criminal law. Check with the local college and universities in your area.

Another resource to consider using is volunteers to assist with crime mapping data and analysis. Undoubtedly, certain public safety personnel will be cautious of this idea. For those who are concerned, I would recommend that they contact agencies such as the Overland Park Police Department in Overland Park, KS, and the San Bernardino County Sheriff's Department in CA, who are currently employing volunteer staff so as to ascertain how the arrangement is working.

For those who prefer to skip the handwritten method and move directly to the computer method, keep in mind the following: To some extent, many of the items mentioned below for the handwritten method will also be required for the computer version. In fact, it would not hurt to have all the items mentioned for both methods. When conducting a search by hand, stationery items, paper products, and writing implements will be required in large quantity. These items should include lined paper and some types of ledger books. If cost is a factor, use standard $8^1/_2$ x 11 paper or legal lined paper and save the money for other expenses. A large quantity of copy machine paper will also be required as well as clear overlays (that can be written on). For writing by hand, have an ample supply of pens and pencils, as well as four different color pens or pencils (and if you are like me, colored pencils with erasers). Also needed are at least two different colored highlighters, staplers and staples, paper clips, and fasteners.

For timeline plotting as recommended in chapter 9, obtain six-month or one-year calendars. Consider the erasable style if the potential for multiple tracks per year exists. As for maps, this will vary with the types of incidents and map sources. Regional, topography, and census maps, as well as photos, as explained in chapter 8, should be included. Other items to consider for the handwritten method are magnifying glasses, rulers, and compasses. All of these items will relate to map plotting and development. Some type of word processor or typewriter will be required for the final report writing. These can be obtained for minimal cost, and in some cases, for free. As other offices and agencies begin to update to computer equipment they will discard most of their word processors or typewriters. Don't forget that once the implements, stationery, and other office products have been obtained, something such as shelves or cabinets will be needed to store it all in.

This process will generate a good deal of paper. Whether by hand or by computer, something will be required to organize the paper trail. File folders, loose leafs, or whatever else is decided, try to choose something that will not be lost with a change of personnel. Develop a file system based on the community, the time period, offender motivation, or any combination of these recommended storage systems. Try to also obtain a briefcase or a carry folder specifically for the reports and related data to be disseminated. This will allow the information disseminated and collected in the field to be located in one collection point. The advantage to the handwritten method is its simplicity. With some form of secure communications and a good quality copy machine, a substantial portion of the analysis process and plotting (maps as well as calendars) can be accomplished in any office space that affords the opportunities mentioned earlier in this chapter.

The first major consideration concerning operating expenses of the computerized method is the compatibility of components. When starting up the process, whatever type of system that can be acquired will be a good start, but if the computer system is of an older donated model, check to see what other components will work with that hardware. For example, an older computer may not work (or interface) with a newer printer. Another example would be that of an older computer's hardware not supporting newer operating systems or newer versions of software. If the computer is offered, try to get everything: the printer, the monitor, software, and all the cables that go with it. Even if it is not as capable as planned, as long as it is compatible with future software and interfaces, this will allow for the process to develop a finished product. That finished product can then be shown to administrators who can then be shown how the process would be improved with upgrades.

When you are able to purchase a newer system, the same idea should apply. Try to get as much as possible for the amount budgeted. A computer will not be of much use unless the results can be printed. Therefore, try to find a printer that will be compatible or be the most efficient with the system specified. Printers will vary as much as the computer in cost and capabilities. Obtain a printer of good quality (preferably a laser or, if possible, a high quality color inkjet printer). Wide carriage color printers are nice, but not required.

As stated earlier, cost for these systems will not be discussed since current prices will be obsolete by the time this book is printed. For specific prices, check with local hardware and software experts within your agency or government. Because of local, state, and federal government purchasing procedures, agencies are required to purchase equipment through a bidding system. Because of this, it is conceivable that the cost will fluctuate drastically when compared to retail pricing. If there is a mechanism to purchase equipment "off the shelf" it is strongly recommended that you shop around. When dealing with hardware (for the computer) it may be worth spending a little extra in price for readily available parts and better service. As for software (the programs and operating systems), "off the shelf" will vary from the ten- to twenty-dollar range to the thousands of dollars. When searching for software, remember to check the discount rack. In fact, it may be the best place to start. This may sound time consuming and unproductive, but remember that time invested in purchasing the proper equipment and training will pay off with good end results. The man-hours in analysis and report production will be cut in half (if not more).

As for hardware, something that is advanced beyond the minimum software requirements would be best. For example, the hardware should have the capacity to easily run the type of software (the database and mapping programs) you require for tracking. Therefore, if a 200 megahertz "Intel" compatible processor is required, go higher. If 10MB of free hard disk space is required, get Gigabites of storage (the higher the better). If 16MB of RAM is required, try to get 32MB or more. If a VGA 256-color monitor is required, be sure the software will run with a higher resolution monitor. If a four-speed CD-ROM drive is required, get double that or even greater speeds. Obtain a 3.5-inch disk drive that will upgrade to allow for a mega-density disk to operate with it as well. Mapping software was discussed in Chapter 8, but continue to check for improved, inexpensive, easy to use map production products. As for a database, be sure it will run with the hardware and software operating system. Preferably the database will allow the data to be incorporated into the mapping software. This refers primarily to GIS software. For example, if there is a potential for the data to be plotted in a GIS format in the future, attempt to identify if the database software language is compatible with the GIS software. This will reduce the entry time necessary; otherwise, the data must be entered into a database form that the GIS software will recognize or be transferred into a different file format. This transfer may cause the potential for lost or corrupted data. If computers are not your thing, it is recommended to get someone with computer knowledge to hook the systems up and then check to see if it all works. In addition, find out if that same person can be called when something does not seem to function properly.

Items nice to have, but not required, are optional. These options are primarily for the computer method, but may have some use with the handwritten method. The first item to consider is a GPS unit. If possible, look toward the hand held units with the capability to upload and download information with the computer employed. Most GPS units will store point locations and plotted trails. This will allow for transcribing if computers are not employed. GPS will be useful in the urban, suburban, and rural MFS scenario. In an urban setting, GPS will allow for pinpointing when involving incident location to the rear of (TRO). When these locations are back alleys, dead-end streets, or other such sites, which are no longer discernible on standard maps, GPS can help provide details to access, vantage points, and travel time. In suburban and rural settings, GPS can give specific addresses to vague locations.

The second option to consider is a digital camera. Digital cameras are good for photographing subjects, crowds, suspicious vehicles, and potential targets. All these images can be digitally displayed on the map or in the final report when using a computer. These units can produce adequate to high-resolution (photographic quality) pictures. Depending on the photo quality required, these units will be cost-prohibitive and will require current computer technology to operate. To remain within a cost-effective range, photo quality may suffer with digital cameras, but simplicity, storage space, and lower overall operating costs will be gained. Digital camera images have begun to be accepted as official crime scene photographs and hence, digital cameras can also be used at crime scene investigations. This acceptance will increase with time and technological advancement. Check with local prosecutors for their opinion on their use.

Still another option is a modem that will allow you to access online Internet services via the phone line. This is one of the few technologies that will help prevent the computer system from becoming obsolete the minute it is plugged in. Internet access adds to a computer system's capabilities in a way that had previously not been realized. Internet service is a tremendous research tool with real or near real-time access to many agencies and systems. Once again, we could spend chapters discussing the future capabilities of this feature, but we only have one (chapter 13) and it will have to share space with many other future concerns and capabilities, so let us move on. Therefore, if a new computer is being considered, be sure it is Internet ready and will operate at or near the current accepted speed with room for future Internet advancement.

One of the last wish list items would be a notebook or laptop computer. These machines can be a great asset to the tracker. They can be used almost anywhere. The latest machines are just as capable as the desktop units. They can be networked and can be upgraded. As stated earlier, these units can be packaged with all the necessary hardware to become a "virtual office" or mobile field unit. With the mobility of your office enhanced in such a way, you can facilitate the actions of multi-agency units working together. The portability of your office will allow for greater input, intelligence gathering, and rapid information dissemination among all personnel. This concept is used regularly by the Suffolk County Pine Barrens Law Enforcement Council. They have frequently established strike forces and joint sting operations in targeted jurisdictions. This portable version of an office with a laptop at the center is considerably more appealing to investigators or trackers who spend an appreciable amount of their time in the field. This could not be easily accomplished with the computer version unless portable hardware such as a laptop is employed.

The limiting factor of laptops is usually the expense. Most are compatible with current stationary devices, such as desktop printers, but if a grant were obtained, it would not hurt to add other portable hardware to the request. Many of the portable printers will print at good to high quality, although they are usually slower in printing speed and higher in cost than desktop models.

Finally, consider topographical prints or digitized topography maps. Topographical prints can assist in identifying the lay of the land. Topographic maps also help identify how the suspects travel (get in and out of fire area), where natural hiding places can be found, and where good observation posts are located. Topographic maps additionally help to explain unusual cluster zones. For example, if some natural barrier separates a cluster zone, it might make it appear as two separate clusters. No matter what method you choose or what space is utilized, keep these basic standards in mind. Hope and shoot for the best, but be prepared to make do with what is at hand. Do not put the process off because all the desired equipment or supplies are not yet available.

Chapter 12

HOW CAN YOU BE SURE
THE PROCESS WORKED?

Now that all the tracking tools are in place, the information has been collected, the track is complete, the reports are written and disseminated, what next? Some may think that the process has been completed and that no further input is required. That would be true if this was a standard crime-mapping unit, but with this process, that would be an error for two reasons. The first and foremost reason is feedback. Feedback is required to see how well or poorly the track worked. This process will, at the very least, always require some fine-tuning. Intelligence and data sources will need re-evaluating. Analysis methods will require upgrades as well as other improvements. The second reason that the process has not yet been completed is that the potential for inaccuracy in the overall process will always be present. As new incidents occur, they will help to improve and focus the track. In other words, the "process" is still in progress. In some cases, numerous new incidents may require a reevaluation of the current track. For these reasons, keep all available lines of communication open. Think of this as a new tool in the toolbox or a new weapon in the arsenal undergoing its shake down or shape-up trials. To assist in receiving feedback, some type of follow-up, critique, or feedback form could be attached to the tracking report. This additional form would be made available to key personnel in the investigation or sent as a request to all involved in the investigation for follow-up on the overall value of the report. In Appendix B of this book are examples of feedback and tracking request forms. Always encourage some type of critique or feedback information and never be afraid of criticism. There will always be ways to improve the process.

In this chapter we will not only cover some of the standard scenarios, but also some that can be considered extraordinary or beyond the normal applications of tracking. Based on the professional disciplines, which were discussed earlier in this book, and their varying requirements for information it is possible that their requirements will dictate the use of this tool for unconventional purposes. Do not automatically limit the disciplines and conditions for which this tool can be used; try to encompass as many as possible. This will keep you active in tracking and help to hone the skills necessary for conducting the process. When you apply the process to other disciplines, the practice of tracking will also help to create additional sources for intelligence gathering. For example, the firefighter and fire officer will be able to use this process in their "size up," that is, when they assess and preplan the strategy and tactics of fire suppression operations. Size-up is an all encompassing term used to improve and explain the preparation for fire ground operations. Consider the fact that the MFS is in the business of placing fire personnel in harm's way. Therefore, firefighters must be prepared to use caution due to the fact that the suspect may escalate to the next level of violence and be seeking to cause bodily injury. Thus, from the firefighter's point of view, the MFS is in an adversarial role and the firefighter must be prepared in the best way possible to combat them. This combat is not primarily focused on catching the MFS, but instead suppressing their handy work, namely finding the best way to put the fire out. Any advantage that can be gained over this adversary in this combat is preferred. By knowing the type, area, and time of a potential incident, the firefighter can use this knowledge to his advantage by being as prepared as possible to take on the opponent. An example of this type of preparedness is found in the Venture County Fire Department in California. This department inputs historical fire incident data into their GIS software to map out a plan for future fire strategies and tactics, as well as prescribed burn operations for wildfire operations. This strategy allows the incident commander to have accurate incident maps and data on the scene within sixty minutes of the request.

Fire suppression personnel will also be able to lookout for suspects at future MFS incidents as well as be alert to protect evidence at the fire scene. Fire officers can also use this process to look inward and to be sure that the problem is not from within their ranks. For this reason, keep fire suppression personnel in the information loop and encourage them to attempt their own tracking. Even though incidents will not always be of a suspicious nature, many will, nonetheless, require some investigation. The following is an outline of the basic steps in the tracking process. We will use "Terrytown" as our example in these steps (the following situation really did occur, but the name of the town has been changed).

Step 1: You receive notification of some odd fire activity in the suburb of Terrytown.

Step 2: Research the electronic intelligence for the previous six months in and around Terrytown and attempt to locate a second data source to confirm any results noted.

Step 3: Refine the six months of data to weed out any non-related incidents such as EMS requests and Automatic Fire Alarms (AFA).

Step 4: Place the refined six months of incident information into a database program format, if that has not already been done.

Step 5: Run the data through the search groups discussed in chapter 6. Note any trends that appear. In our example, one trend appears consistent.

Step 6: Analyze the information and compare it with your second information source. You may find the type of fire, the time that fires are set, and the locations where they are set are all consistent. In our example, the location appears to center at the Terrytown Railroad Station. The incidents occurred near the very end of rush hour traffic and the types of incidents have all been rubbish fires. Days are fairly consistent in that fires occur on weekdays with no weekends noted. No other incidents within the community fall within this pattern. When the railroad's police database is compared, these types of incidents are only being noted at the Terrytown Station.

Step 7: Map the incidents. In our example, the map places all incidents approximately 600' from the local fire station.

Step 8: Plot the calendar. A quick calendar plot does not indicate anything except that no incidents were noted during the federal holidays. The local fire chief has begun to notice the faces appearing at the incidents. It is at this point that the chief recognizes one of the same faces at the majority of the incidents. The face happens to belong to one of his volunteer firefighters.

Step 9: Decide what to do. Usually you have two choices. The first is to formally complete the track and monitor the activity. In our example, this would be based on the fact that the level of violence has not escalated and that no damage has been reported to the station. An abbreviated report could also be generated to the concerned agencies to follow the situation. The second course of action would be to further investigate the incidents, based on the fact that only three of the four required traits have been verified with one more (the days) as similar, but not specific, and the noting of one subject from a reliable HI asset. In this scenario, no one direction could be considered favorable over the other. Either direction would hopefully bring the investigation to the proper conclusion.

In our example of Terrytown, further research would reveal the following: that the personnel employed to clean the railroad station had retired approximately six months ago and had not been replaced to this point. Due to this situation, refuse would begin to collect at the station. Furthermore, due to the fact that no smoking is allowed on the trains, the commuters regularly attempt to get their last smoke in before making the long trip into the big city. Because there are no proper cigarette receptacles at the train station, any place will do, including the corner or below the platform where trash would accumulate. Any fires that started due to this cause, while commuters were at the station, were quickly extinguished and would go unreported. Unfortunately, the fires that started after the commuters had taken the train were not quickly extinguished. These fires did require FD response, but were not serious enough to require further investigation until they became frequent. As for the potential suspect, he was returning from his swing shift job and would not have had enough time to set the fires before the end of his shift, although he was close enough to the community to have his personal fire radio activate when an alarm went out. He was also not close enough to make the first responding vehicle. This is important because of the proximity of the fire station to the railroad station. Even if the subject were to drive faster, this would only allow him to make the second or third responding apparatus. With the return of cleaning personnel, the fire incidents ended. If these incidents had occurred in a larger community, the fire chief could have consulted his department fire prevention officer. The prevention officer could introduce some simple cost effective methods for eliminating the fire problem. For example, in this case a simple solution would be for the railroad station to obtain some proper cigarette receptacles.

An additional use that the fire prevention officer can use this process for is to help identify target areas for anti-arson and fire prevention programs. These programs could inform the residents of particular areas about how to protect their property from becoming a target. (This concept of informing the community was developed by the Federal Church Fire Arson Task Force in the mid 1990s. This led to Congressional and Presidential action such as the National Arson Prevention Initiative and the [Church] Arson Risk Assessment Guide and Checklist.) Posters offering rewards for information could be placed in the same area, which may ultimately lead to an arrest due to residents being reminded to be on the lookout. In scenarios described in this chapter, we will try to follow the basic step-by-step process as it is outlined in the flow chart found in appendix A.

One of the most effective ways to apply tracking is to involve local patrol officers into the process. These personnel are the eyes and ears of the target area. Involving these personnel has the potential to inject new motivation into the investigation. Many patrol officers take pride in their patrol area or the community they protect. Any constant criminal activity in their sector will reflect on them and their professionalism. Many times they welcome the opportunity to be involved with the investigation and will work hard to crack the case. This opportunity to incorporate highly motivated professionals should not be overlooked. As mentioned earlier in this book, patrol officers can use the process to establish patrol routes and to take note of vehicles, subjects, and activity. They can also use fire incident track information to assist in interviewing witnesses and area residents about specific activities, types of subjects, their means of transportation, and so on. If a tracking report contains the description of a possible suspect, it can be the basis for stopping, interviewing, and frisking unknown subjects who meet that description if they are in the fire area. In some situations the tracking report, combined with additional intelligence, could be used for probable cause. However, before pursuing this course of action, it would be recommended to consult the prosecutor in the jurisdiction for clarification and operating procedures. At the very least, the information developed from this process will make patrol officers more aware and better attuned to noting information related to the area and the incidents.

For the crime analysts and crime mappers, a tracking report will have some specific advantages. The tracking process allows the data that has been collected for years to be used as it was intended, that is, for fighting crime. The greatest advantage to the crime mapper is the ability to help identify what the fire investigation requirements are and how they differ from other investigations. These requirements include the need to weed out and refine actual suspicious incidents from non-suspicious, as well as MFA calls from good intent calls. The tracking report introduces the crime mapper to new data and intelligence sources. By applying the process and producing the report, the crime mapper will be better prepared to interact with fire investigators on their search requirements and target analysis, as well as expressing these goals in the report for dissemination.

Some agencies are limited not only in manpower, but also equipment and technology due to budget restraints. By addressing concerns and goals in the report, common limitations can be overcome with other limited assets. As brought out in the first scenario, if the data was produced into a report without consulting the fire investigator or fire prevention personnel, many man-hours could have been wasted and the data could wind up being sent to inappropriate agencies. Instead, forwarding the report to the appropriate public safety agency will allow the data to be utilized in their follow up inspections, future preplans, and specification designs.

As for fire investigators, this process will have the greatest use. Once completed, the investigator should apply all the information mentioned before and any additional items not mentioned to the MFS investigation. Attempts should be made to use all the data in the most advantageous way possible to create what is known as the "home field advantage".

Exactly what do I mean by "home field advantage"? Since I learned how to play football as a young boy, I have loved the game. I watch it anytime the opportunity arises. No matter what level of football game is played—high school, college or professional—it is a well known fact that the team that is playing in their own field has the advantage over the other team. This home field advantage can provide tremendous momentum to the home team. If you have ever been to a big college football game or basketball game, you know what I am talking about. The audible support of the hometown fans can make it impossible for the visiting team's quarterback to call out the cadence or for the visiting team's defense to call out defensive assignments; but home field advantage is more than just the enthusiasm of the fans. For instance, Pennsylvania State University plays outside in Happy Valley where, at least until the late 1990s, the field was known for the wind that would do some very unusual things. This wind would play havoc with the visiting team's kicking game if they were not prepared. This type of anomaly is what defines home field advantage.

Let's examine another discipline, that of the military. In military terms, the same concept is called "dominating the battlefield". Napoleon Bonapart was known for choosing the ground where his army would do battle with their enemies. He identified the area that offered the best advantage for his strategy and tactics.

Now let's apply these same ideas to the MFS. As with other serial offenders, most MFSs are motivated by their need to seek power–power over others or power to dominate persons within a geographic area. By setting fires, the MFSs will believe that they dominate, control, or have power over others. This supposed power comes in many forms. The NCAVC has noted the characteristics of harassment and intimidation as key motivations when occupied structures or properties are targeted. A case with these characteristics occurred in the county where I am employed. The suspects torched cars of individuals they once went to high school with, never confronting the individuals. Their objective was to intimidate and dominate the battlefield, so to speak. The investigator must use the tracking process and report as tools or weapons to take the home field advantage away from the suspects and dominate the MFSs instead.

At this point it can be said that great football teams overcome their opponent's home field advantage. Every effort is made to nullify and place the home team and their fans on a downward spiral, taking away the momentum from the home team and using it to place them in retreat. The investigator must adopt this same concept. Use the process to help dominate the field and set the terms and tempo; know that true power is based on information. The process is the method of gathering, sorting, analyzing, and disseminating the information. By taking the power away from the MFSs (without them knowing it) the unsub can be forced into an area of operation on the investigation's terms. Just as overcoming home field advantage is not easily accomplished in a football game, neither will it be easy for the tracker to take the home field advantage away from a MFS. Nevertheless, what separates great teams from the rest is the ability to take the home field advantage. The same must be said for the investigators. A word of caution at this point is to not allow the success of one case to cause you to become overconfident just because one track leads to an arrest. Neither let an unsuccessful track discourage you. Continue to rely on the facts. Never forget that technology dictates tactics; continue to update the tracking process with new technology as it becomes available, affordable, and reliable. Use it to dominate the battlefield.

Now that we have covered how to use the process and discussed how the process can work with non-MFS cases, we will continue to look at some MFS scenarios. The following scenario has been developed to show trackers how to put the tracking tools to use and bring all of the professions mentioned earlier together via the tracking report in order to dominate the field. This is not to say that the NCAVC should not be consulted on this subject. This is once again only a view from the perfect world scenario with the subject being a typical MFS (if that is possible). Once the process has been completed and a track has been established, the next step is to identify subjects or suspects. With the aid of patrol and investigative officers, a subject is identified. After reviewing the intelligence on the subject, the investigators feel confident that this is a viable suspect. The problem of the investigators is that they are not confident that the suspect will admit to the fire setting without substantial proof; therefore, a plan of action is required.

Plan of Action

1. Once the active fire areas are established, attempt to identify the most recent cluster, then identify the oldest or least active (possibly ground zero). Once completed, establish surveillance and move to step 3. If this is not possible, then move to step 2.
2. If only one cluster has been established, then identify an unoccupied property proximal to the suspect's single cluster or active area. This property will depend on the level of escalation; it can be open land with tires or trash, a truck loaded with bails of hay, or an abandoned house. Whatever type of property it is, try to choose one that is owned by your agency or government. Check with the tax assessor's office, department of real estate, or whatever agency handles evictions in your jurisdiction. If your government owns the property, then there will be no question of trespassing or the pressing of charges after the act. Choose property with conditions conducive to setting up surveillance.
3. Set up surveillance on the inactive cluster or the staged target and get the word out to the suspect that the property is ripe for fire. This can be as easy as breaking windows on

the property or posting eviction signs. MFSs are known to "go cruising at night" to survey the hunting ground or they may tend to travel past potential targets during their daily routine. If the MFS is traveling by personal vehicle, then it is probable that there is a high frequency of travel within the area of activity. With this idea in mind, the suspect has probably been summoned or stopped for some type of traffic violation in his or her vehicle in or around the active fire area. Do not be surprised to find out that your suspect frequents the area and has a valid excuse for being there. Getting the word out may require using a CI to mention it in conversation with the subject. During one case, where the suspect was a tow truck operator, the plan was to place abandoned vehicles in front of the county-owned target property. The subject would be called to remove the car and the eviction officer would make a point of mentioning that this property was vacant. Local laws covering baiting and luring subjects will dictate the way the notification is accomplished. Placing for sale signs on the property or placing an obviously abandoned car nearby are two other ways you can let the suspect know that the property is an easy hit. This can prompt him or her to act.

4. With the exception of the surveillance site, flood all other active fire areas with high profile patrols, arson awareness hand outs, and anti-arson and arson reward posters. This is usually easily accomplished without calling in enormous amounts of overtime personnel. Simply use the tracking report to identify probable times and days. Then, bring in patrols from less active areas or utilize special patrols such as K9 units or auxiliary officers. In this effort you should utilize law enforcement, fire personnel, and whatever assets are available to the investigation. If necessary, schedule units based on the tracking report for twenty- to thirty-minute intervals to maintain coverage of the patrol areas. This type of rotation has led to positive results. For example, in the city of Detroit on the days leading up to Halloween there has regularly been increased intentional fire activity. The night before Halloween is referred to as "Devil's Night" due to the amount of mischief and violence that occurs. Since

the late 1980s, patrols have been increased on "Devil's Night" in target areas based on historical activity and intelligence gathered. This has led to a dramatic decline in "Devil's Night" fire activity. Remember to inform all patrols to steer clear of the surveillance site. Continue the patrols based on the pattern reports and maintain them until the subject strikes or moves on. Have patrol units write down the numbers of license plates of vehicles stopped or parked in active fire areas. This may be difficult in crowded urban areas, but remember David Berkowitz, the "Son of Sam" was caught based on being given a ticket for being double parked on a crowded street. This placed him in the area at the time of two murders.

5. If possible, push the suspect's buttons. In other words, if you know what will trigger him or her, use it. The objective is to set the MFS off and cause him or her to act. Use your pattern information. Pick the time and day most likely for activity and trigger the subject just prior to that time. This may meet with resistance. Some will not be comfortable with this since the outcome might possibly be pushing the suspect to escalate the violence. There is a potential for this, but a pattern search should reveal what level of violence the subject is at. If the suspect has escalated to the level where the potential for violence is that severe, then a tighter surveillance may be required, preferably on the suspect.

6. The last step may be the most difficult of all, patience. Continue your patrols, canvass the active fire area again, and wait. Talk to the neighbors; remind them to be on the lookout for unusual activity and remind them who to contact. When they ask what you are doing, tell them that you are doing everything possible and when they ask where you are, tell them that you are everywhere. Remember, if no fires are reported, you are still accomplishing two goals–stopping the fires and protecting the neighborhood. Do not become discouraged.

The next scenario involves one of the most difficult tracks to accomplish, eco-terrorists operating in a long range or wide geographic area. These MFSs can be almost impossible to identify without a little help or, in some cases, unless a dose of luck is interjected. The scenario that we will explore will involve a big dose of luck. If you recall back in chapter 2, when we discussed eco-terrorists, I mentioned that they are usually organized and difficult to catch. The one thing we have going for us is that eco-terrorists (as well as most other terrorists) love to take credit for their handy work. If they did not, there would be little sense in their actions. Hence, the MFS for this scenario will many times actually identify their work for us. Most terrorists wish to ultimately make statements that will persuade others to take up their cause and stop whatever the terrorists see as wrong or contrary to their way of thinking. Fortunately, the fact that the terrorists do take credit for their work will ultimately help us to track all their activity and remove any doubt related to which incidents were and were not the work of the MFS. Therefore, we can get right down to the business of tracking.

The first step is to identify all of the target incidents related to the MFS eco-terrorist(s). In this scenario, the MFS have not only burned buildings under construction, but also painted slogans and defaced certain corporate facilities. The defacing of the buildings occurred prior to the fire setting (a sure sign of escalating violence) and was directed at certain fast food corporate offices. The fast food corporate offices attacked were a vast distance away from the known fire activity (over twenty miles). The fire setting activities were targeted over three separate communities separated by at least 5 miles, but still within the one regional geographic area. This regional area encompasses the central northern section of the county in question (we will call it Peconic County and the region encompasses the Villages of Soundview, Middleton, and Lakeside in the Township of Thompson). This area is approximately 40 square miles in size (See Map 12-1).

The targets have been housing developments and condominiums with multiple residences burned per incident. Slogans that were found on the corporate offices were also found on the residences burned, although the terrorists have stated that all new construction that threatens natural wildlands will be attacked. The only area signaled out is in the regional area discussed earlier. No activity has been noted in other sections of Peconic County or the neighboring counties where much large construction sights are noted. This is especially noteworthy based on the fact that many other housing developments under construction infringe on more sensitive wildlands than those in the targeted region. Additional intelligence (via Federal Law Enforcement sources) identified that the fire setting techniques follow the textbook methods outlined in the eco-terrorist manifesto manual, found at their Internet website.

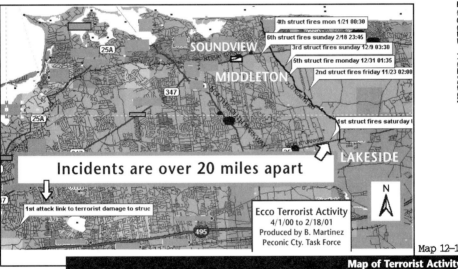

Map 12–1:

Map of Terrorist Activity

Mapping and plotting the incidents also identify that most of the fire setting incident sites can be accessed via one secondary road (Old North Rd.) that runs through all the affected communities. The first fire incident occurred off a primary road (Middle Turnpike) that the identified secondary road (Old North Rd.) crossed over. After the initial fire incidents (in Lakeside) the activity moves to the northern most community (Soundview) in the identified regional area. The activity then proceeds south through Middleton and back toward the original incident in Lakeside.

When all of the incidents (fires and defacing) are plotted on a calendar and checked for days of the week, it is noted that the activity has primarily occurred on weekends (Fridays, Saturdays, and Sundays). It was also noted that some of the incidents occurred on federal holidays and seasonal holidays. The incidents span a time-frame of ten months and appear to have begun last spring, with activity increasing toward the months of December through February. The times identified show that most incidents occurred in the late evening hours on Friday and Saturdays, and the early morning hours on Sundays. The exceptions to the Sunday incidents were when they were preceded by a holiday in which case some incidents occurred on Sunday evenings.

So what do we have? Let us begin with target assessment and attempt to identify similarities. Our analysis has identified two clusters; the first was at the fast food corporate office, and the second was in the regional area within the Township of Thompson. What do they have in common? In reference to the geographic area, the similarities are minimal. The corporate offices are located in an office park and the residential construction projects are on former farmland. The types of targets attacked are consistent within each targeted area. The distance between the two clusters suggests some form of transportation. Although there is public transport available in and around the office park, the first cluster, there was no regular public transportation available near the second cluster area (the residential developments). More than likely, the MFS terrorists used some private form of transport when travailing between the two separate clusters. The fact that the second cluster area was concentrated to one regional area only suggests further research into its political, geographic, and demographic features. The basic results from this type of search are that the second cluster area is a predominately middle to upper class affluent suburban area with good community awareness and moderate crime activity. As for the political subdivisions, it is protected by one police district and three fire districts. It has one school district, and is located within one township.

As for the corporate offices (the first cluster), they also fall within an area of upper middle-class residences, with many single residences being built on large parcels of property. The homes are set back off of main streets and are surrounded by wooded areas. The same police district, as in the second cluster, protects this area, but not the same fire districts, nor is it in the same school districts or township. The corporate target in the first cluster (the fast food chain) is a well known and established conservative company. They hire predominately individuals new to the job market or first time employees.

Based on the mapping information we can conclude that the MFS terrorists know the geographic area that encompasses the second cluster. The same conclusion cannot be reached for the area surrounding the first cluster or the area between the two clusters; this is based on the fact that the MFS did not attack any targets outside the second cluster that were much easier to attack. The potential targets noted outside the second cluster were easier to access, threatened larger wildlands, and the structures being built were usually larger.

Based on the facts identified, our report conclusions should focus on the following: The subject(s) work and or live within the regional area where the fire incidents occurred (the second cluster). Look for future incidents to occur on weekends, with special attention to holiday weekends. The timeframes to watch will be between 2000 to 0000 and 0000 to 0400. Probable targets will be residential construction sites in and around the second cluster (the villages of Soundview, Middleton, and Lakeside). Vehicles frequenting Old North Road (the secondary Rd. noted in the plotted map) during the days and times specified should be identified. This report should be disseminated to all local patrol officers in the designated target area, all investigators handling incidents or related incidents, and all available special patrols, as well as state and federal agencies who are assisting (See Table 12-1).

Multiple Fire Setters

PECONIC COUNTY ARSON TASK FORCE
CONFIDENTIAL REPORT

PAGE 1 OF PAGES 3

REPORT ORIGINATED BY W. Horst	REPORT # 001	COMPLAINT NO. 00-0202		DATE OF THIS REPORT 2/20/01
LOCATION Thompson	SCHOOL DISTRICT Greater Middleton	COMMUNITY Soundview, Middleton	FIRE DEPT. Soundview, Middleton Lakeside	TYPE OF INCIDENTS auto, debris/ brush, dump, struct

AGENCIES NOTIFIED
County Police, County Fire Marshal, State Police, FBI, BATF,
Tri-State Joint Terrorism Task Force.

The analyses of nine incidents occurring in the Townships of Thompson and Hartland, from 4/1/00 through 2/18/01 have revealed the following patterns.

NOTE: The information that follows is not a profile but a study of the facts surrounding these incidents. These facts include the time, date, day of week and location of the incidents. The information in this report has been gathered from two independent data sources (The County Fire Rescue Communications Log and the County Police Incident Report Database).

FACTS:
A pattern of fire and malicious mischief incidents has occurred in two separate and distinct clusters that are separated by a considerable distance.

All these incidents have been linked to eco-terrorists who have taken credit for these attacks.

The clusters have been broken into two areas. The first cluster is at the office complex in Hartland Township. These targets involve the defacing of property owned by fast food chain corporations on three separate occasions.

The second cluster is in Thompson Township, where the targets were housing developments under construction in the communities of Soundview, Middleton, and Lakeside.

All the incidents in these communities have occurred in developments directly on or adjacent to Old North Road. This road appears to be the common denominator in this cluster.

The days that these incidents occurred are consistent in both clusters and have only deviated during federal and state recognized holidays.

The times have also been consistent in both clusters, regardless of the day.

The incidents that occurred in the second cluster (Thompson) involved multiple fires being set at each site by using incinerary devices with time delay fuses (for further details on the type of devices contact the Joint Terrorist Task Force).

The activity began in April of 2000, at the first cluster (Hartland) and then moved to the second cluster (Thompson) in June of 2000.

The activity continues at the second cluster with no activity noted at the first cluster.

A check of surrounding towns and counties has not identified any additional activity. All incidents have been focused on these two clusters.

Table 12–1:
Fire Tracking Report for Eco-terrorist Scenario

PECONIC COUNTY ARSON TASK FORCE CONFIDENTIAL REPORT

PAGE 2 OF PAGES 3

REPORT ORIGINATED BY W. Horst		REPORT # 001	COMPLAINT NO. 00-0202		DATE OF THIS REPORT 2/20/01
LOCATION Thompson	SCHOOL DISTRICT Greater Middleton	COMMUNITY Soundview, Middleton	FIRE DEPT. Soundview, Middleton Lakeside		TYPE OF INCIDENTS auto, debris/brush, dump, struct

AGENCIES NOTIFIED County Police, County Fire Marshal, State Police, FBI, BATF, Tri-State Joint Terrorism Task Force.

CONCLUSIONS:

Due to the distances involved and the transportation of accelerants and incendiary devices, the unknown subject(s) will probably use some form of private transportation. Look for activity to continue to occur in the communities of Soundview, Middleton, and Lakeside in the Township of Thompson. The tracking analysis projects activity to occur on the following days.

Fridays - Look for incidents to occur between the hours of 2200 to 2359. Activity should focus on the areas around Old North Road. Probable targets are residential housing projects under construction.

Saturdays - Look for incidents to occur between the hours of 0000 to 0400 and again at 2200 to 2359. Activity should focus on the areas around Old North Road. Probable targets are residential housing projects under construction.

Sundays - Look for incidents to occur between the hours of 0000 to 0400 and again at 2200 to 2359 when Monday is a recognized federal or state holiday. Activity should focus on the areas around Old North Road. Probable targets are residential housing projects under construction.

Monday - Look for incidents to occur between the hours of 2400 to 0400 when Monday is a recognized federal or state holiday. Activity should focus on the areas around Old North Road. Probable targets are residential housing projects under construction.

For additional information or to report information contact FM Parrett at 555-5555.

Table 12–1:
Continued . . .

Multiple Fire Setters

PECONIC COUNTY ARSON TASK FORCE
CONFIDENTIAL REPORT

PAGE 3 OF PAGES 3

REPORT ORIGINATED BY W. Horst	REPORT # 001	COMPLAINT NO. 00-0202	DATE OF THIS REPORT 2/20/01

LOCATION Thompson	SCHOOL DISTRICT Greater Middleton	COMMUNITY Soundview, Middleton	FIRE DEPT. Soundview, Middleton Lakeside	TYPE OF INCIDENTS auto, debris/ brush, dump, struct

AGENCIES NOTIFIED
County Police, County Fire Marshal, State Police, FBI, BATF, Tri-State Joint Terrorism Task Force.

4th struct fires mon 1/21 00:30

6th struct fires sunday 2/18 23:45

3rd struct fires sunday 12/9 03:30

5th struct fire monday 12/31 01:35

2nd struct fires friday 11/23 02:00

1st struct fires saturday 6/9 01:

SOUNDVIEW

MIDDLETON

LAKESIDE

Ecco Terrorist Activity
4/1/00 to 2/18/01
Produced by B. Martin
Peconic Cty. Task Force

1 miles

N

Table 12–1:
Continued . . .

Two weeks after the reports are distributed (and on a holiday weekend) the MFS terrorists strike again. This time the target is a construction company with offices near the second fire incident in the village of Soundview. This incident reveals some important facts. First, that the track was correct on the time, days, and probable location of attack. The MFS did not break the pattern, in spite of the aggressive patrols occurring in the targeted area. Based on that fact, it would be safe to assume that the track was successful in preventing an attack on the primary targets, those being the residences under construction. We can also assume that our prediction that the MFS terrorists live or work in the area is correct because they struck an area that they are most comfortable with instead of venturing out to less protected, similar, and easier targets. With some added insight from seasoned investigators and based on the facts developed from the tracking report, it is assumed that the subjects are young adults or teenagers, who attend the local high school or are first year local college students. The subject(s) own or operate a vehicle (most likely a car). The subject(s) were once employed by the fast food restaurant targeted in the first cluster area. The subject(s) do not travel with family members during holidays or school closures. Furthermore, the subject(s) do not travel often outside the geographically targeted areas.

Based on these conclusions, investigators reach out to CI (confidential informants) in and around the school district that covers the second cluster zone. Quickly, subjects who are known to be sympathetic to the eco-terrorist cause are identified. The subjects' residences and the types of vehicles that they drive are identified. When investigators check with the officers who patrol the targeted area, it is noted that the officers identified one of the subject's vehicles on Old North Road in the hours preceding the most recent fire incident at the construction sight. Now that the subject noted in the area before the most recent event is considered a suspect, surveillance is established on the suspect and patrols in the targeted area are notably reduced. During the surveillance, the suspect is spotted with other subjects known for their environmental protests at the Thompson Town Zoning Board meetings. In less than two weeks, the suspect is seen with another identified subject, purchasing less than two gallons of gasoline at a service station in the target area and proceeding toward a residential construction sight off

Old North Road. Patrol units are called in and the suspect is stopped for a traffic violation, making a "U" turn over double yellow lines. The officer notes, in plain sight, devices that are consistent with the type used to accelerate the suspicious fires noted in the target area. The suspects are brought in for questioning. This last scenario helps to show the synergistic effects of good tracking results, decent intelligence, and cooperation among investigators.

The next scenario we will discuss is a reverse fire incident track. This would be done after a suspect has been identified or arrested. This type of track is useful when building a case for trial or interviewing the suspect. It may also be a good way to start working with the process. My first cases involved a reverse fire incident track. This method allowed me to better understand future MFS techniques and targeting methods. If you have the time, I would recommend that you try it with some closed MFS cases.

For this case, let us assume that you are with the Fire Tracking Unit of the Municipal Fire Department's Bureau of Fire Prevention and Investigation. A state parole officer has requested your assistance with a recently released convicted arsonist. The parole officer feels that the subject (who will be referred to as "Ralph") is still a threat to society, but does not have the proof he needs to revoke his parole. The parole officer would like your help in proving his case. The reason for his concern is based on Ralph's history. The facts are that Ralph was arrested for setting a minor fire in a government facility (the County Court and Municipal Building). Ralph was also seen setting fire to a bulletin board outside the local unemployment office in the morning hours before the office officially opened around 0800. Ralph was convicted and sentenced to 5 to 10 years. Unfortunately, the state correctional facility that Ralph was sent to was overcrowded and did not need another arsonist running around its minimal security facility. Therefore, Ralph was transferred to a half-way house, where he was identified as a model inmate that had shown remarkable remorse for his actions. The supervisor's glowing report allowed Ralph to get an interview with the parole board, and consequently, an early release. Ralph served 18 months of his expected 5-year term. After meeting Ralph on his first parolee interview, the parole officer felt the subject had been rushed out. This type of quick release was known to occur with many subjects who were considered non-violent felons and that is especially true with arsonists. Additionally, it is not uncommon for management at correction facilities to fear an arsonist's potential for setting fires in their facilities. Based on this idea, most facility managers will attempt to get rid of them as soon as possible. The parole officer feels that Ralph either has, or soon will, revert back to his old tricks. What can you do to help him stop Ralph from becoming a threat?

First, we will start by reviewing Ralph's past for incidents that either support or refute the parole officer's thoughts. To accomplish this goal, we will review the subject's history of known residences, previous employment, driver's license and record, and vehicle ownership. The search reveals that Ralph has lived in the village of Stony Port for all of his adult life. He is married and has two sons. Ralph has been unemployed for approximately six years. The parole officer supplies us with Ralph's known criminal history and that of the subjects living with him. Ralph has a history of minor offenses including public drinking and motor vehicle infractions. Ralph's sons have similar violations, as well as domestic violence charges. They all reside at 17 Cosmos Drive in Stony Port. A search of police calls at Cosmos Drive reveals numerous neighbor complaints for items such as loud music, rubbish, and domestic disturbances. The motor vehicle history shows that Ralph's driving privileges had been revoked for the past five years and that Ralph has not owned a vehicle for the past four years.

The next step is to research the fire activity in and around Ralph's residence and the site in and around the area that led to Ralph's arson conviction. It is recommended to research back at least two years from Ralph's conviction or as far back as records will allow (five years is preferred in this case). At this point, you would employ the fire incident track process to weed out any of the incidents that do not concern the track (such as AFA, EMS, and MVA calls). Follow through with the standard analysis, calendar and map plotting. The incident history and tracking process reveals that for the past five years, fire activity involving rubbish, abandoned vehicles, and abandoned structures has been consistent in Stony Port.

The activity appears to occur during the late evening hours every second and fourth Thursday and Friday of the month. The only break in the activity appears to occur during the 22-month period immediately after Ralph's arrest and his incarceration. When the activity is plotted over a map of Stony Port, it identifies two clusters. One cluster focuses around a four-block area of Main Street and the other focuses on a six-block area around Cosmos Drive. The fire history search of the site where the known fire setting occurred reveals that three similar fires occurred over the same 5-year period. At least two of the incidents appear to be suspicious in nature, with no investigation of the third. These incidents all occurred at the Municipal Building on similar days and at similar times, although not exactly the same as the incident for which Ralph was convicted. The incidents have enough similarity with Ralph's known incident to meet the criteria for a pattern. The fires all occurred before the beginning of business hours and were similar in nature (ignition of available material in the hallway or main corridor of the structure). The incidents were separated by a ten to twelve month timeframe that spans the same five-year period of the Stony Port incidents. The next step is to formulate the information into conclusions and write a Fire Tracking Report (See Table 12-2).

METROPOLITAN FIRE DEPARTMENT
BUREAU OF FIRE PREVENTION & INVESTIGATION

CONFIDENTIAL NOTIFICATION OF MULTIPLE FIRE SETTER ACTIVITY

To: All Metro Firefighters - Stony Port Div., State Parole Office, State PD - Troop C, Metro PD 6th Prct. - Patrol, Det. and K9 Divisions.

Date: 6/12/04   Authority: C. Ember

NOTE: The information that follows is not a profile but a study of the facts surrounding these incidents. These facts include the time, date, day of week, and location of the incidents. The information in this report has been gathered from two independent data sources (The Metro Fire Dept. Communications Log and the Metro Police Dept. Incident Report Database).

This report is based on fire incidents that occurred over the past five years in the community of Stony Port and at the county court and municipal building. The analysis of fire activity at the two noted areas has revealed the following:

Table 12–2:
Fire Tracking Report for Parolee

A pattern of fire activity involving woodlands, abandoned autos, structures, and trash receptacles has been occurring in the areas of Main Street and Cosmos Dr. in the community of Stony Port on and off for the past five years. It appears that the fire setting activity has resumed recently around Main Street and Stony Port Road. This activity occurs primarily on Thursday and Friday nights. The subject sets fires using available material and will target any of the above mentioned sites.

CONCLUSION:
Units should focus on a four block area around the corner of Main Street and Stony Port Road on Thursday and Friday evenings and early morning hours (between 1800 to 2359 and 0000 to 0200) the following day. Targets will include wooded areas, trash receptacles (of all sizes), and abandoned automobiles and structures. Attention should also be paid to Cosmos Drive for future unknown fire activity. Note any subjects wondering in and around these areas during the specified times or active fire incidents.

See the attached map for further details.

Cosmos Dr.

Stony Port Active Fire Area
Prepared by P. Brennan 6/10/04

N

Main St. and
Stony Port Rd.

Table 12-2:

Continued . . .

Once completed, you can forward the report to the parole officer, fire investigators who handled any previous cases, and patrol
officers who operate in the identified cluster areas. Also forward the

report to your command with recommendations to alert fire suppression crews who also operate in the known cluster areas. This will encourage the parole officer to contact fire investigators, fire suppression personnel, and patrol officers in Stony Port. The inquiries reveal that Ralph was known for frequenting drinking establishments in and around the Main Street cluster. When the parole officer speaks with bartenders and establishment owners about the days identified in the track, he is informed that the second and fourth Thursdays of the month are the dates that the subject receives his disability checks. Ralph has been receiving the checks for the past six years. More research by the parole officer into the criminal history of Ralph's sons puts both of them in court two of the days that fires occurred in the municipal building. Court records show that Ralph's sons paid fines on the days of the previous fire incidents and that the payment was made with a check from Ralph. It would appear that Ralph was at the municipal building on the days in question to pay his sons' fines. He more than likely arrived by public transportation. The only available public bus from Stony Port to the municipal building arrived before 0800. Based on this information, the tracking report, and HI (human intelligence) received along Main Street, the parole officer was able to set up a loose surveillance of Ralph by using normal patrol and fire suppression assets (See Map 12-2).

The parole officer also informed fire investigators and special patrol units to be aware of the day and time patterns of fire activity and pay special attention to subjects in the area during any future fire events. Fire personnel identified all large trash containers, abandoned cars, and structures in the area between Main Street and Cosmos Drive in Stony Port (See map 12-3). This information was forwarded to all agencies that received the fire tracking report. Within two months, Ralph appeared to be up to his old tricks again.

Recommended Patrol Area for Identifying Stony Port Fire Setting Suspect
Prepared by P. Brennan 6/10/04

Corner of Main St. and
Stony Port Rd. in Stony Port

N

Map 12–2:

This was produced from the New York State GIS Clearinghouse,
"LI South Shore Project", www.nysgis.state.ny.us/raster.htm

Metropolitan Fire
Department

Probable targets for fire activity in community of Stony Port
Prepared by P. Brennan 6/10/04

Abandoned Structur

Abandoned Autos

Trash bins

Brush

Main St.

Stony Port Rd.

N

Map 12–3:

This was produced from the New York State GIS Clearinghouse,
"LI South Shore Project", www.nysgis.state.ny.us/raster.htm

Although he was not identified setting a shed fire, a patrol officer responded to Ralph's home upon notification of the fire. Upon arriving home, Ralph spotted the patrol unit and immediately began walking in the opposite direction. The patrol officer pulled up next to Ralph and told him, "we have been waiting for you." Ralph immediately asked, "Who is we?" The patrol officer told him that some people were waiting to speak to him back at the station. The parole officer and fire investigator were notified and met Ralph at the police station. During the interview, they explained how they were waiting for Ralph and that they had evidence that he set the shed fire and many others. It was apparent that Ralph had been drinking and he was amazed that they knew so much. Ralph gave a statement on how he set the fires and why. Although Ralph moved up and down the scale of violence when setting fires it was only a matter of time before Ralph tied one on and attempted to set an occupied structure ablaze.

In this last scenario, it should be pointed out that there is a potential to ruffle a few feathers or uncover mistakes in previous investigations. When backtracking, there is a good potential for uncovering information that will place individuals from other agencies or similar professions in a bad light. Whether they are missed leads, wrong determinations of fire causes, or unidentified fire patterns, they should not persuade you to stop or cover them up. Remember, rotating shifts, bad weather, and a whole host of other reasons can get in the way of an investigation, especially when it involves fire incidents. Mistakes do happen and no one is at fault. Keep your eye on the goal and things will tend to all work out in the end. As you may have figured out, all of these scenarios have been based on actual cases, although varied to protect the innocent. The conclusions in these scenarios may have differed from the original cases, but the concepts covered are all based on fact.

The last scenario we will discuss is one that is based completely on fact and is a testament to good investigation and cooperation all around. This scenario should be used as a model for the last identified use of the tracking process, that of preventing further incidents. Yes, the tracking process can also be applied to prevent future activity, thus allowing the track to be part of a proactive investigation. This scenario is based on a case that occurred in

Suffolk County during the worst drought conditions in recent history. In the summer of 1999, the potential environmental conditions were considered to be worse than the drought conditions seen in 1995. That particular year saw the loss of half a dozen homes to wildfires, as well as numerous evacuations in the communities of Rocky Point, Ridge, Westhampton Beach, and Eastport. The largest of these fires effectively closed Sunrise Highway (State Route 27) for over five days and was renamed the Sunrise Fire. During the height of the summer season, a fire on the Southern Fork of Long Island caused by this fire resulted in substantial economic loss. To help prevent a repeat of this tragedy, considerable efforts were made by public safety and land management personnel. Two of those efforts involved the most modern technology available. The first involved the monitoring of fire activity in the protected lands of the Pine Barrens with the Fire Incident Tracking System. The second involved the incorporating of the Fire Weather Index as a fire investigative tool. The Fire Weather Index has been studied since before the 1995 fires. With the assistance of Dr. Robert Parish of the US Fish and Wildlife and their computerized weather monitoring ability, the daily index could be identified. All the concerned agencies were able to receive and disseminate the daily index to all fire departments as well as prevention and investigation personnel. These two tools would play a key role in the coming summer months of 1999. Within days of the monitoring effort, suspicious fire activity was noted along the border of two fire districts. The districts in question were historically active during the summer season and were located in the center of the Pine Barrens region. Based on this fact, the activities of the two districts were researched and plotted and fire investigators were notified. Within hours, the fire investigators confirmed that the fires were "man made" and probably of a suspicious nature. Due to the dry weather conditions and the confirmed "man-made" fire activity, all information was shared between all investigators and analysts immediately. The fire incident track identified that the fire activity was not simply limited to only two fire districts. The track revealed activity on a scale that had not yet been seen. These facts were quickly shared with all public safety officials in the targeted areas, as well as with land management officials. At the same time environmental conditions became critical.

The Fire Weather Index had now been rated as "High" for 16 days and would go to "Extreme" for the next 5 days. To put this in perspective, during the 1998 and 2000 seasons combined, there were a total of only six days when conditions were rated as "High" with no registered "Extreme" days occurring in either year. This sobering fact, combined with the fire incident track, led to two proactive events. The first was the restriction of all recreational camp and barbecue fires in all town, county and state parks, as well as restricted access to all protected lands within the Pine Barrens region. This was not an easy decision for elected officials to make during the peak time of the season. Fortunately, by identifying that the fires had been "man-made" and that the environmental conditions were dangerous, the elected officials had the evidence necessary to enact such drastic measures. The second event was the establishment and deployment of public safety officials to probable target areas based on the fire incident track results. These patrols consisted of any available town, county, and state public safety agencies. Their primary assignment was to operate in targeted areas, thereby producing a high profile presence with the ability to enforce the restrictions. The final result was that although one of the busiest wildfire seasons was recorded in both Nassau and Suffolk Counties in 1999, the majority of incidents occurred on unprotected lands, primarily in western Suffolk County. At the same time, the Pine Barrens Protected Lands recorded one of the quietest wildfire seasons on record. This example shows how good intelligence coupled with current technology can be applied to develop proactive measures to prevent known criminal activity.

Chapter 13

THE FUTURE OF TRACKING AND THE PROCESS

We have arrived at the final chapter of this book. For me, this has been the most cautious chapter that I have written. The reasons why are simple. First, the one sure way to outdate a publication is to involve a discussion of future technologies. This idea is due to the fact that whatever is mentioned as "the future technology" will become the current technology by the time of publication, and history by the time it is in print for a year. The second reason why this discussion is not a good idea is because whatever technology is decided on as holding the key to the future, it will still be considered an unproved and untested system. In some cases, this "key to the future" will be outperformed by some completely new development in a competing technology. The only safe way to consider technology of the future is to base it on a concept that is so futuristic in thinking that no current developments exist to perform the function. This type of thinking would be moving us from the realm of technological advancement to the world of theory and science fiction. Thinking in this way will not help the people currently working in the field for whom this book was intended. Instead, it would be for someone who will be starting out twenty or thirty years from now. Somehow we must stay within the reality of current technological advancements and the future. We must try to identify technology that will continue to be user friendly enough to accomplish the daily requirements of the process. By being user friendly, it will also survive the advancements of unproven developments.

What must be avoided is a system that can only be operated by computer programmers who are not accessible to the personnel in the field. More than thirty years ago in the 1970s, Dr. David Icove developed an idea that was decades ahead of its user-friendly availability. That development, computerized fire tracking, has been a key component to this process. The only problem back then was that the technology and operability needed were not easily accessible. Dr. Icove assumed that fire tracking would be in full swing by the 1980s. To borrow a line from my favorite sportscaster, Warner Wolfe, "If you took Dr. Icove and the ARPS system (Arson Pattern Recognition System) to be operational by 1990, you lost." The reality, unfortunately, was not as we all would have hoped. This is not to say that ARPS cannot be accomplished; instead, it only shows that Dr. Icove was just ahead of his time. In fact, the learning curve continues and hopefully can move on after all, just as that same sportscaster further states, "The Future is now." This means that you do not wait and hope for what can potentially be. Instead, employ what is at hand and what can currently be gotten to accomplish the goal. This is the same idea for consumer sales.

As discussed in chapter 8, the KISS principle (Keep It Simple, Stupid) should apply when future concepts and purchases are considered. Start with an inexpensive basic design that can be added to and upgraded. This is the concept used by many corporate giants such as Harley Davidson Motorcycle. They offer a basic model (in this case the 883 Sportster at the lowest price possible) to allow the rider the opportunity to appreciate the Harley experience and get people riding. This allows people to add on or upgrade to the bigger, more powerful, and more expensive bikes at their own pace. Too many of the GIS programs try to pack everything into one package instead of breaking the product down into separate components that will allow users to get started and appreciate the usefulness. That usefulness will turn to necessity as personnel in the field apply new ideas and concepts that allow them to become more efficient with each new addition. If GIS manufacturers do not realize this, they will be left behind by less expensive computer mapping software, which is very user friendly, capable of adding on components and offers inexpensive upgrades. For the non-computer programmer, the future is now because these less expensive systems allow them to get started. These software manufacturers already realize that the cheaper and easier the software is the more personnel in this field will utilize it. Consequently, this will increase software sales and allow the manufacturer to continue to offer greater improvements at discounted rates.

In the past, the expense or cost of new technology would have been the primary reason why future technology would not be adopted. To a certain extent that continues to be a factor, but when current technology is applied to future concepts, the expense is not the primary issue. As explained in chapter 3, those who are willing to do research and write the grant proposals necessary can attain the funding available. The cost of current technology can be overcome by other means. Sure, new technology is expensive, but old technology is sitting on the shelf waiting to be integrated with a little modern development. In addition, other advancements have made systems previously thought unthinkable, affordable.

In keeping with the idea that the future is now, we will spend the remainder of this book discussing future concepts that are achievable now with the technology that does exist. For example, while writing this book, my wife and I were blessed with the birth of our first child. During the first months of his arrival, we were able to watch and listen to our son in his crib in almost complete darkness from any room of the house with a wireless television monitor. This was a godsend; we could do the wash, clean the kitchen, or sleep in our own bed and be able to monitor our son at all times. Why bring this up? Simply exchange the word monitor with surveillance and imagine how this simple inexpensive technology could be adapted to monitor remote locations. The cost of this system at the time of publication is less than two hundred dollars. Connect a video recorder and this can be an unmanned surveillance system for the monitoring of target hazards.

Future concepts in surveillance do not end with that example. Another idea would be to utilize the interstate closed circuit video monitoring systems of many state highway departments. These systems monitor interstate and major highway traffic day and night in all weather conditions. They can be found along most major highways in densely populated (urban sprawl) areas. The systems work on a closed circuit network that allows the operator to pick and choose the camera with the best view. The majority of these cameras are able to rotate and move up and down. Some can also zoom in and out. Many law enforcement and security agencies have invested in this technology. In doing so, they have developed their own (expensive) network. Rather than spend all the funds on one or two cameras, why not use the funds to tie into the 20 or 30 cameras already in place. Many of these monitors are only used by the transportation agencies during peak traffic hours. This would leave them available during the remainder of the day and nighttime hours. These cameras could be used in place of, or as a complement to, the existing surveillance equipment. You may have to work to gain the cooperation between varying departments, governments, and personalities, but nothing that cannot be overcome when the value is realized.

Another future concept involves current satellite technology. The National Geophysical Data Center (NGDC) under the National Oceanographic and Atmospheric Administration (NOAA) operates the sensor data archive for all government weather satellites both defensive and civilian. These satellites will vary in government ownership with the majority being operated by the US Air Force DMSP (Defense Meteorological Satellite Program) at Offutt Air Force Base. These satellites can be accessed via the Internet at www.ngdc. noaa.gov/dmsp/. The reason this is of interest to trackers is that these satellites can detect fires at night. This technology can spot fire activity throughout the world and can be used in situations where large tracks of land would be difficult to put under surveillance. By using satellites as surveillance tools, other assets can be used for other investigation. This satellite technology can also be used to help pinpoint the origin of fires that are not immediately detected by other means. By pinpointing the exact origin of a fire in wildland fire situations, the entry and exit points can more easily be established. This will help identify methods of travel, time of ignition, and surveillance positions. The growth of private satellite and aerial imagery will also play a role here, especially when one factors in the advancements in resolution. It may be possible to show a photo of not only the trails and tracks, but also the before and after conditions of the area. This would allow for other activity and/or the offender's casing of the area to be noted. Most commercial satellites and aerial photographic companies keep archives of the images taken over the years. This is also true for many federal, state, and local government agencies. If the MFS is believed to be an extremist-motivated offender, there is a good possibility that they tested their devices and techniques to ensure success. By incorporating the calendar plot with the viewing of archived aerial photos with current photos of uninhabited areas near the active fire areas, it may be possible to identify the unsub's testing site. Once this test site is identified, the area can be checked for other activity, evidence, and clues that may help identify the unsub. This same technique can be used with known suspects to show what activity has occurred on their property. Add to this the fact that varying types of digitized aerial photos can identify vegetation growth, new construction, and ground disturbances. This is accomplished with infrared, thermal, and radar scanning imagery. You can check with an aerial photo supplier to see if this type of imagery has been taken.

Another advancement that will assist with future concepts in surveillance is in the area of tracking by wireless phone. Although it is illegal in some states to listen to phone conversations on wireless phones, it would be legal to track someone by their phone. As wireless phones become increasingly popular, phones will become cheaper and more prevalent. With the government now requiring GPS to be installed in all new phones, the capability for it to be tracked will exist within every phone. This will have a domino effect on GPS technology that allows the GPS devices to keep pace with the wireless technology. As the phones become smaller and more capable, so too will GPS continue to get smaller in size with better receiving capability and greater accuracy. As an MFA pattern emerges, a track can be established. Knowing the track, a trap can be set. As the calls are placed, agencies will be able to activate GPS devices within the phone, allowing the tracking of the subject's movement.

The majority of future concepts, which use current technology, exist primarily on the computer front. The use of linkage analysis for the purposes of crime solving is one such advancement. In Washington State, the ability to integrate technology with past case studies has developed into HITS (Homicide Investigation and Tracking System). HITS was developed for WSAGO (Washington State Attorney General's Office). HITS relies on multiple databases found primarily within the criminal justice system of each state and on ANN (the declassified Artificial Neural Network). The databases include the corrections and probation systems as well as special task forces such as gang, vice, and sex crime units. HITS can also access the State Department of Motor Vehicles. HITS first collects all the addresses of the incidents and then identifies similar MOs and the unsub's signature at the various crime scenes. Any tips such as vehicles seen in the area are then added to supplement the information. When an incident is noted, the records are checked for similar incidents occurring in the area. The data is then cross-referenced with known offenders or parolees who frequent the crime area. A list is then developed to see which former offenders meet the MO and profile of the cases. These names and pertinent details are then prioritized onto a map by a GIS program.

The ANN actually has the ability to learn from former mistakes and adapt. It also studies closed cases to learn what similarities can be applied. Although this system is big, expensive, and only operates with known felons, it contains the key concepts of success. It is based on a sound idea and can be adapted to work with current systems and HI. For example, to adapt this concept to MFS tracking, simply start working with other criminal justice agencies and their databases to achieve the same goals. It will not be as fast or as infallible as HITS, but as time and analysis continue, improvements can be made, efficiency will increase, and adaptability will improve. By incorporating profiling techniques, the picture will become clearer of when to employ this linkage process. For example, by allowing the information developed in the FITS (fire incident tracking report) to be analyzed by a profiler, hopefully the unsub can be categorized as a parolee who would be old enough to drive or a juvenile who would not show up in a HITS search. The use of ANN technology or standard human networking could then weed out those incidents that are not part of the MFS pattern thus refining the track as discussed in chapter 7. In choosing which course of action to follow and which offender to focus on (such as parolees, youthful offenders, or gang members), the ANN technology would have the obvious advantage by increasing accuracy and decreasing analysis time required.

A similar system currently in use that will adapt to future concepts is the work done in the area of geographic profiling. Geographic profiling or geoforensic profiling incorporates standard psychiatric profiling techniques with GIS technology to help identify how different criminals utilize geographic space, that is, their hunting grounds. The process focuses extensively on the questions of "Where?" and "What?" Specifically, where did the incident occur and what was the land use? By playing on the detailed knowledge of certain areas that individuals may have and their relationship to those areas and including spatial relationship to the crime sites, then criminal activity can then be established. Doctor Kim Rossmo, formerly of the Vancouver Police Department in Canada, has developed this type of profiling into a computer software system called "Orion". This software uses GIS technology and mathematical formulas to produce graphs and three-dimensional maps that show probable sites of offenders' residences and/or places of employment. Adding more cases can then refine the geographic profile and produce a crime scenario that represents the likelihood of future criminal activity by a serial offender. In the hands of a trained profiler, this can be a powerful tool. This computer program utilizes many of the ideas of the MFS tracking process and incorporates additional concepts, such as known offenders, to help produce a probability as to the identity of the unsubs. Once again, incorporating this technology with FITS reports could produce a specific offender residence or neighborhood, allowing for greater success in surveillance and apprehension.

As noted in the previous chapters, the future of tracking is not limited to criminal activity. In fact, in the United Kingdom, the Kent Fire Brigade has proven that limited technology can help keep you ahead of the curve. The fire brigade has incorporated some GIS technology to identify the fire problems in particular areas and who in those specific areas are at the greatest risk. A good example of this concept would be to employ the fire protection assessment programs now being done in the US. These programs identify high hazard areas in wildland/urban interface zones. Once these areas are identified, prevention programs are developed for the target audience—that is, the owners of residences that have been built in heavily wooded areas usually in remote areas. This target audience is approached at events such as civic meetings, festivals, or school functions. During the brush fire season, we set up manned tables at parks or festivals in the communities that are at the highest risk. We attend the annual open house at one of the largest federal facilities we have, BNL (Brookhaven National Lab). BNL sits on the western edge of the Pine Barrens. Fire prevention experts (who can inform many more people than could be reached by going door to door) staff these tables. These types of projects refer to fire prevention programs for fire patterns that are identified to be accidental in nature. Examples would include improper campfires or misuse of all terrain vehicles that have been known to spark brush fires.

Other types of fire activity identified, such as fire setter activity, will be targeted with arson prevention programs and law enforcement notifications. The use of fire protection assessment to identify high hazard areas in wildland/urban interface can be applied to MFS tracks to help refine data and conduct historical comparisons to current activity.

By combining many or all of the technologies discussed earlier, the MFS tracking process could achieve a capability not yet realized. "When?", "where?", and "why?" will be left to future professionals, but the potential here is great. As for the question of "How?" we will consider the most likely and the most promising current technology that may some day lead to a potentially secure Internet or intranet computer web site. The Internet is the most capable current technology for almost all of the future concepts of tracking discussed. Besides being one of the best tools to solve current issues surrounding the MFS tracking process, it is also remains one of the most cost-effective. For example, the Internet can help solve many of the problems raised in earlier chapters, such as creating and disseminating maps, as was discussed in chapter 8 (See the appendices for a list of Internet sites). This can be accomplished by accessing various sites over the Internet. Many states currently offer maps, aerial photos, and GIS base data on their websites. In many cases, this information can be downloaded and manipulated to create basic fire activity maps. Check with the state government web sites in your area for details. In NY State one of the most useful sites is the NY State GIS Clearinghouse at www.nysgis.state.ny.us. This site contains not only information related to GIS data, training, and reference help, but also contains links to other useful web sites. All of the information contained in these web sites is available to the general public at no cost.

Take note that due to the nature of the track information, any tracking network setup will have to have limited access so as to prevent the subjects being tracked from becoming alerted. How a secure website is accomplished is not within my area of knowledge, although current technology can accomplish this task. Having said this, the fact remains that the Internet has the potential to be the most helpful facilitator in the sharing of information. The Internet can take products such as "AIMS" (the Arson Information Management System) and incorporate it into a national level to disseminate information on all types of fires and fire setters. This would not be a replacement to the NCIC (National Crime Information Center) but an addition to it, based on the facts involving the uniqueness of fire incidents and the difficulties involved in investigating them.

Some of these future concepts discussed can already be accessed via the Internet. When all of these concepts are in place and can be accessed, then a society or association of MFS trackers will be able to not only share information, but also attempt to assist each other with tracks, scenarios, and conclusions. This will help enhance individual capabilities and general professional knowledge. As more and more agencies move into the 21st century, these capabilities will continue to increase. The use of the Internet with tracking will allow even more investigators to discuss cases, perhaps even tracking MFSs nationwide, eventually catching that professional torch mentioned in chapter 2. This information dissemination via the Internet will not be limited only to stopping criminal activity, but fire prevention officers will also be able to share tracks for future safety measures. If this does not come about on a national scale, there may still be a site for sharing and exchanging information at a relatively secure private enterprise website. Unfortunately, these private enterprise websites would not be free and may operate for a hefty price that creates exclusion. Hopefully, as more agencies move to operating "online" the collecting of information, as discussed in the perfect world in chapter 6, will move closer to reality with the ability for investigators to share information, data reports, and maps online in real time or near real time.

Other areas requiring further research for future concepts would be the environmental issues discussed in chapter 7, such as the phases of the moon and their effect on suspects. There have been studies done by mental health experts on the effect of moon phases on humans, but no real evidence on fire activity exists. Included in environmental studies should be further work on which climate conditions have the greatest effects on MFS. Do more MFSs exist in any particular climate or region than other regions? All of these questions require greater research.

In order for this process, or for that matter, any form of tracking to continue to work and grow, there is one issue that will have to be addressed. That issue is that individual investigators will have to get past the concept of wanting to make the big arrest. Investigators must get past wanting to be the one to take "the walk" or the one giving the press conference. "The walk" is the term used to explain the photo opportunity for the news media to see the suspect led, in handcuffs, by the arresting officer to an awaiting vehicle on their way to be arraigned. The investigator must see the bigger picture if this process is going to succeed.

Generally speaking, crime mappers or trackers will not be making the arrests. Remember that the goals of this process include preventing additional fires and tracking down those who are suspected of setting the fires. This may mean that uniformed officers on regular patrol will make the arrest or the arrest will be made by some other law enforcement agency with the FITS report. Other jurisdictions may take credit for the arrest or your work. Do not be surprised when this occurs. Remember, by sharing the FITS reports and keeping all informed, including your superiors, the word will get out and individuals will receive the credit due them. I personally have made two of the arrests of the 21 MFS and multiple MFA cases that I have been involved with. That does not mean that the process failed; in fact, investigators now routinely request that a FITS report be generated based on intelligence that they have received. That, in and of itself, is progress.

By now it should be apparent that this process is by no means perfect. Based on what has been discussed, it is fully operational with the current technology and information sources available. That is not to say that there is not room for improvement; this is the reason for the writing of this book. It is important to tap the minds of those fellow investigators, fire officials, crime mappers, and GIS students currently working in the field to increase the productivity of the tracking process. Hopefully, this will encourage debate and in turn increase the capabilities on the subject of tracking. The introduction of new technology with fresh new ideas will move the process along even more. What does the future hold for this process? Hopefully a great deal. I know we may be getting way ahead of ourselves, but hopefully, through online Internet chat groups or working groups within established organizations such as the IAAI, Arson Task Forces, NCIC, NFA, and Criminal Justice students, the discussion will continue. The tracking process will continue to be refined, the number of experts and those knowledgeable in the field will increase, and ultimately lives and property will be saved.

Appendices

APPENDIX A

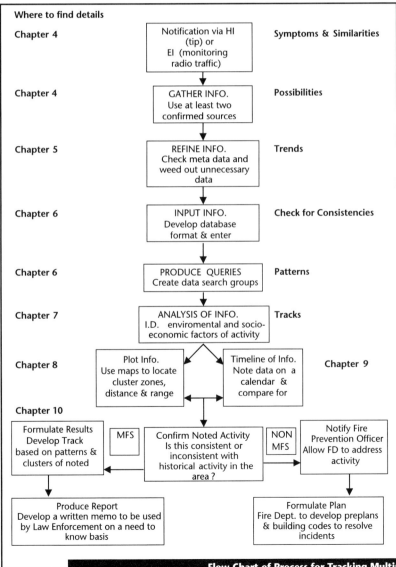

Where to find details

Chapter 4 — Notification via HI (tip) or EI (monitoring radio traffic) — Symptoms & Similarities

Chapter 4 — GATHER INFO. Use at least two confirmed sources — Possibilities

Chapter 5 — REFINE INFO. Check meta data and weed out unnecessary data — Trends

Chapter 6 — INPUT INFO. Develop database format & enter — Check for Consistencies

Chapter 6 — PRODUCE QUERIES Create data search groups — Patterns

Chapter 7 — ANALYSIS OF INFO. I.D. enviromental and socio-economic factors of activity — Tracks

Chapter 8 — Plot Info. Use maps to locate cluster zones, distance & range

Timeline of Info. Note data on a calendar & compare for — Chapter 9

Chapter 10

Formulate Results Develop Track based on patterns & clusters of noted

MFS

Confirm Noted Activity Is this consistent or inconsistent with historical activity in the area ?

NON MFS

Notify Fire Prevention Officer Allow FD to address activity

Produce Report Develop a written memo to be used by Law Enforcement on a need to know basis

Formulate Plan Fire Dept. to develop preplans & building codes to resolve incidents

Appendix A:

Flow Chart of Process for Tracking Multiple Fire Setters

APPENDIX B

The Ten Commandments of Crime Analysis

By Christopher W. Bruce
Crime Analyst for the Cambridge, MA,
Police Department

I developed the Ten Commandments of Crime Analysis for a presentation to the International Association of Crime Analysts in the Fall of 1999. The presentation was also given to the Massachusetts Association of Crime Analysts at the 11.12.99 meeting. The purpose of the Ten Commandments is to identify the fundamental principles of crime analysis and to thus help define the profession. I welcome your comments, anecdotes, and objections at **cbruce@ps.ci.cambridge.ma.us**. *I don't mind if anyone copies or otherwise reproduces this list, but since I intend to eventually use the Ten Commandments in a publication, please attribute them to me somewhere in your document, e-mail, or presentation.*

1. Thy Task is Crime Analysis. Thou Shalt Have No Other Tasks Before It.

Crime analysts tend to become technological wizards in a relatively short time, given the number of computer applications vital to modern crime analysis. This computer proficiency tends to make a mark on the other members of your department, and eventually you find yourself mired in requests to develop a database for the Internal Affairs Unit, to crunch citation numbers for the Traffic Unit, and to help an investigator design a flyer for his upcoming housewarming party! On the other hand, maybe your unit has simply developed an overall "reputation for competency", so that when anything important needs doing, the command staff tends to "give it to crime analysis". Whatever the case, you must develop techniques to put your primary task—crime analysis—at the top of your list every day.

2. Thou Shalt Read Thy Department's Crime Reports Every Day.

Too many crime analysts try to find trends and crime patterns by looking up the information in their records management system (RMS) or computer-aided dispatch (CAD) system. This approach presents many problems with timeliness, accuracy, and information sufficiency. You want your information to be accurate and timely. Furthermore, you want to have access to the full text of every crime report so you can correctly identify modus operandi and categorization. To this end, you will find no substitute for reading (every day) copies of all crime and arrest reports taken by your department.

3. Thou Shalt Track and Control Thine Own Information.

Another problem with records management systems: they generally do not have fields that allow you to enter information vital to crime analysis. Such information may include point of entry (for burglaries), type of premises, categorization or classification, whether you have identified the crime as part of a pattern or series, and many other modus operandi factors. If you want to track this information across a period of time—and trust us, you do—you will probably find your RMS inadequate. You should develop your own means for recording crime patterns, and crime conducive to patterns, in your jurisdiction. Methods for this include matrices, spreadsheets, and—probably the most ideal—customized databases.

4. Honor Thine Patrol Officers and Investigators.

Remember, the job of a crime analyst involves identifying crime patterns and trends so that the patrol and investigative divisions can develop strategies and allocate resources. You are a tool for their use, not the other way around. If you become aloof from or hostile to your patrol and investigative divisions, you will probably fail as a crime analysis unit. Crime analysis is impossible without accurate information, and patrol officers and investigators are fonts of information. Unfortunately, much of this information is undocumented. You may have identified and analyzed a pattern of robberies in the South Central district, but only Sergeant Jones knows that he saw a pack of suspicious looking kids hanging out there the night before last. You will need to develop a good rapport with your officers and investigators in order to facilitate the exchange of this type of information.

5. Thou Shalt Never Present Statistics (or Maps) Alone.

"There are lies, damned lies, and statistics" is a quote variously attributed to Samuel Clemens, Winston Churchill, Benjamin Disraeli, Karl Marx, and Theodore Roosevelt. In many cases, this is true. Statistics presented alone, with no comparison or context, are like nuclear power: they can be used for good or evil. Never present statistics by themselves. Statistics are indicators; your job as a crime analyst is to interpret them. Never say "There were thirty housebreaks in the Old Port Neighborhood last month" and leave it at that. Statistics must be comparative, descriptive, and accompanied by qualitative analysis: "There were thirty housebreaks in the Old Port Neighborhood last month. This is up twenty percent from the previous month and thirty-five percent from the same period last year. The average neighborhood in the city averages fifteen housebreaks per month. Of the thirty housebreaks last month, ten were crude, kick-in-the-door-and-steal-the-VCR jobs (which is usual), but twenty were sophisticated breaks in which alarm systems were circumvented and expensive oil paintings and oriental rugs were stolen. This is an unusual modus operandi for the city, and we therefore attribute the increase in burglaries to a new professional burglary ring that is at work in the city."

6. Thou Shalt Know Thy Jurisdiction from One End Unto the Other.

The first task of any new crime analyst should be to get to know his or her city or town. If you are an officer-analyst with several years of patrol under your belt, this probably won't be a problem for you. Civilian analysts, however, may know little or nothing about the city when they start their jobs. This can result in some comical blunders. You don't want to report a "major pattern" of shoplifting at 100 Main Street if 100 Main Street is a megaplex mall where shoplifting occurs every day. Very quickly, you will want to learn the street layouts, the major parks and public areas, the major commercial areas, the neighborhood boundaries, the ethnic enclaves, the economic situation of each area of the city, the locations of public housing projects, and generally where people live, where they work, and where they spend their free time in your jurisdiction. As you grow as a crime analyst, however, you will want to know more. Your ability to analyze patterns and to recommend strategies will be much greater if you have personally visited the crime "hot spots," patronized the commercial areas, and driven through the depressed residential areas.

7. Thou Shalt Not Stop Crime Analysis at Thy Jurisdiction's Borders.

There's an old legend about King Arthur. One day, Merlin turned him into a goose and let him fly across England. As Arthur sailed through the air, he surveyed the landscape below him. He saw mountains, and rivers, and plains, and cities, but he realized to his astonishment that he couldn't tell where one country began and the other one ended. There were no lines on the ground—as there were on maps—to mark the political geography that men held so dear. Work with maps long enough, you'll start to unconsciously think of the areas outside your jurisdiction's borders as blank white tundra. These imaginary lines need to be unimagined for at least two reasons. First, while your police administrators may value the difference between one side of a border and another, you can be sure that your criminals don't care. If you have a crime series that potentially crosses another jurisdiction's boundaries, you'll want to get together with your counterpart over there to see if they're experiencing it too. You can both help each other out with information. Many departments are experimenting with regional information sharing systems—but until a regional "system" comes to your jurisdiction, a few simple telephone calls once or twice a week can have impressive results.

8. Who, What, When, Where, How, and Why Are Thy Children. Thou Shalt Not Favor One Over the Others.

Proper analysis of crime patterns and trends involves careful consideration of all factors. New technologies tend to give emphasis to certain factors over others—the advent of "crime mapping", in particular, tends to overemphasize the "where" factor. Departments with advanced GIS systems often rely on their mapping to identify all crime patterns, even though many patterns do not show themselves in neat clusters—finding these patterns usually involves a careful reading of the modus operandi, or the "how" factor (see Commandment #2). The key to crime analysis, like all things in life, is balance. Both effective identification and effective description of a crime pattern requires the intelligent consideration of factors like victim and suspect characteristics (who), the type of crime (what), the time, day, and date (when), the modus operandi (how), the cause or offender motivation (why), and of course the location, type of premises, and geography (where). Make sure all factors are considered in your analysis.

9. Remember Thy Community and Keep It Holy.

In the end, your direct supervisor is not your boss, nor is it your bureau head, nor is it your commissioner or chief. You ultimately work for the people who live, work, and play in your jurisdiction, and your job—like the job of the police department as a whole—is to make their lives safer. Anything that accomplishes this goal (e.g., daily tactical crime analysis) should be your A1 priority. Anything that is not related to this goal (e.g., administrative reports) should take a lower priority. Anything that is antithetical to this goal (e.g., blowing off pattern analysis for a week, caving in to departmental encouragement [to] "hedge" the numbers) should not be done at all. Try to remember that there are dozens—or perhaps hundreds or thousands—of people who have not been victimized because of your work. If you identify and thus help your department stop a pattern after four incidents instead of a dozen, that's eight people who weren't burglarized, robbed, or vandalized because of you. They'll never know it, and you'll never know them, but never forget that they exist.

10. Thou Shalt Not Covet Thy Neighbor's Neural Network.

If you're a new crime analysis unit, sooner or later you're going to attend a regional or national conference on crime analysis, where you'll discover that the police department up the street is using neural networks, is engaged in data mining, and is conducting raster mapping. You'll realize with shock that you don't even know what these terms *mean*. Stop the feelings of inadequacy and inferiority before they start and remember this: 75 percent of the benefit of crime analysis is achieved through the basic tasks of reading reports, looking for crime problems, and issuing bulletins to your department. Neural networks and other advanced technologies are helpful for certain departments, but then, we've met crime analysts who talk about neural networks but who couldn't identify a serial rapist if he was spelling his name across the city, and we've met brilliant crime analysts who couldn't find the power switch on a computer. Advanced technologies, like basic skills, are only tools with which you perform your essential duty—identifying, analyzing, and reporting crime patterns and trends. Your superiority as a crime analyst will depend on how well you do your job, not on what tools you use to do it.

APPENDIX C

Internet Resources for Tracking Multiple Fire Setters

The following is a list of useful Internet sites that may be of interest to you or your associates for training, assistance with problems, related certifications, and future developments in tracking. I would like to give credit to Susan C. Wernicke, a crime intelligence analyst with the Overland Park (KS) Police Department. Susan compiled the majority of this information and deserves to be recognized for it. Please note that some sites may appear under more then one heading. This is based on the fact that some sites offer more then one service.

College, University, and Certification Programs

• UEE Certificate Program in Crime and Intelligence Analysis

The Education Certificate in Crime and Intelligence Analysis site explains practical instruction in crime analysis and criminal investigation.

http://www.csufextension.org/Classes/Certificate/
CertificateClasses.asp?GN=3011&GV=2&LID=

• Crime Analysis and Crime Graduate Certificate Program

This site covers the college's academic programs, financial aid for students, resources faculty directory, staff directory, discussion forums, continuing education research office, and computer support.

http://www.cohpa.ucf.edu/crim.jus/map

- Certificate in Crime and Intelligence Analysis @ UCR Extension

  The Certificate in Crime and Intelligence Analysis @ UCR Extension works in cooperation with the State of California Department of Justice. It discusses the Certificate in Crime and Intelligence Analysis programs that are designed for inservice training.

  http://www.unex.ucr.edu/certificates/CIA.html

Commercial Products

- Enviromental Systems Research Institute (ESRI)

  GIS software company that offers products such as Arcview.

  http://www.esri.com

- Enviromental Criminology Research Inc. (ECRI)

  Geographic profiling products.

  http://www.ecricanada.com/geopro/index.html

- ESRI and The Omega Group announce extended development of crime analysis GIS technology.

  http://www.giscafe.com

- MapInfo

  GIS software company

  http://www.mapinfo.com

- i2 Visualization and Analysis Home Page

  Offers data-mining products such as the Analyst's Notebook, a visualization tool for businesses. Download products and white papers.

  http://www.i2group.com/

- AGC: Web Resources
  GROUP CENTER FOR CRIME and INTELLIGENCE ANALYSIS TRAINING

  Crime Analysis Resources, Corona Solutions,
  www.coronasolutions.com

  http://www.alphagroupcenter.com

- Garland Police Department Crime Analysis

  Crime analysis unit

  http://www.ci.garland.tx.us/police/Gpdca2.htm

- Crime analysis

  Enhance the value of your crime analysis data with interactive maps and crime reports.

  http://www.geotrac.com/products_services/gis/crime_analysis.html

- Geographic Information

  http://www.geotrac.com/products_services/gis/

- Crime Analysis and Beat Book Disclaimer

  The Crime Analysis and Beat Book Disclaimer offers crime analysis application and community policing beat book application. These projects were supported by Award No. 97-IJ-CX-K020, awarded by the National Institute of Justice, Office of Justice Programs.

  http://www.arcdataonline.com/industries/lawenforce/index.html

- Southwest Software Solutions, Inc.

  Southwest Software Solutions, Inc. offer CrimeSTATs, an integrated Crime Analysis module that offers agencies a more in-depth analysis of statistics. Reports are based on monthly and yearly totals of specific Part I or Part II Crimes and Work.

  http://www.crimestat.com/

- InfoMart Area Crime Analysis

  Offers applicant screening services, systems, software, area crime analysis.

  info@ers.infomart-usa.com

  http://www.infomart-usa.com/

- Crime Analysis Support System (CASS)

  CASS is a file management system that provides file creation and report flexibility for any public safety application. In addition, CASS increases the report ability of all CISCO software.

  http://www.cisco-ps.com/cass.html

- Violent Crime Linkage Analysis System (ViCLAS)

  This is the site of the Violent Crime Linkage Analysis System (ViCLAS) of the Violent Crime Analysis Branch at RCMP.

  http://www.rcmp-grc.gc.ca/html/viclas-e.htm

- Infoshare Case Study

  Brent Crime Analysis Case Study;
  Brent Crime Analysis Index Home Page.

  http://www.infoshare.ltd.uk/brent.htm

- CrimeSTAT's, Inc.

  This site offers an integrated crime analysis module that offers agencies the ability to view crime trends over a 10-year period for investigative analysis.

  http://www.crimestat.com/

Mapping

- MapInfo

  This site offers sample applications of GIS Mapping for Law Enforcement.

  http://www.mapinfo.com/industry/government/application_areas/index.cfm

- CAPSE: Crime Mapping and Analysis

  Information pertaining to crime mapping and analysis. Crime related software for law enforcement on the Internet.

  http://www.geo.hunter.cuny.edu/index2.html

- Portland Crime Analysis: Mapping and Charting

  Neighborhood Crime Research by Oregon Professional Microsystems; Professional Microsystems Introduction to Crime Research; Glossary Caveat.

  http://www.worldstar.com/~carltown/crmemenu.htm

- Mapping & Crime Analysis Bibliography Page 2

  The final goal of the project is to review the literature in the fields of Crime Analysis, GIS, Epidermiology, and other social sciences to find useful applications and/or services.

  http://www.geo.hunter.cuny.edu/index2.html

- SCAS Home Page

  The Spatial Crime Analysis System introduction is an ArcView-based GIS application designed to allow police departments to perform SOP functions.

  http://www.usdoj.gov/criminal/

- Crime Analysis Application Extension for ArcView GIS 3.1

  This application is designed to provide easy-to-use tools for geographic crime analysis, data management, mapping, and reporting.

  http://www.esri.com/industries/

- Geographic Information System Criminal Justice Links

  Analysis and research, geographic, criminal justice. I would like to invite any agency that is working toward the development of criminal justice geographic information systems.

  http://www.co.pinellas.fl.us/bcc/juscoord/egis.htm

- GIS/LIS
  This is an improved approach to crime pattern analysis.
  http://www.odyssey.maine.edu/gisweb/spatdb/gis-lis/gi94001.html

- NLECTC Rocky Mountain - Crime Mapping and Analysis Program
  Mountain Region Crime Mapping and Analysis Program.
  http://www.nlectc.org/cmap/

- Crime Mapping and Analysis - Crime
  Missing Persons - Other crime mapping and analysis
  http://www.crimeresearch.org.za/

- Geographic Information Science and Crime Analysis
  http://www.ucgis.org/apps_white/crime.html

- GIS in Health and Crime Analysis
  The role of GIS in crime analysis.
  http://www.geog.leeds.ac.uk/research/presentations/97-2/

- Crime Spots: GIS Applications in Crime Analysis
  Discusses GIS Applications in crime analysis, suppositions about
  policing, and crime rate drops.
  http://www.ualberta.ca/

- GIS Use Increasing for Law Enforcement and Crime Analysis
  This site showcases crime analysis in police departments and
  some examples of crime analysis using MapInfo.
  http://www.tetrad.com/new/crime.html

Associations and Organizations

- International Association of Crime Analysts - IACA.net
  Crime mapping research organization
  http://www.iaca.net/

- Cambridge, MA, Police Department
  Site of the Ten Commandments of Crime Mapping
  http://www.ci.cambridge.ma.us/~CPD

- Crime Mapping Research Center (CMRC)
  Government-run site for exploring crime mapping.
  http://www.ojp.usdoj.gov/nij/maps

- Crimemap listserv
  Chat room for crime mappers.
  listproc@aspensys.com

- Massachusetts Association of Crime Analysts
  Crime mapping data and training.
  www.macrimeanalysts.com

- Police Executive Research Forum
  Research organization.
  http://www.policeforum.org

- Police Foundation
  Crime mapping research organization.
  http://www.policefoundation.org

- The International Association of Chiefs of Police Website
  http://www.theiacp.org/

- The International Association of Law Enforcement Intelligence Analysts Website
  http://www.ialeia.org/

- Justice Information Center homepage
  http://www.ncjrs.org/

- Law Enforcement Analysis Website (to join list serv)
  http://www.inteltec.com/leanalyst/newadd.shtml

Articles/Publications/Speeches

- CRIME SEARCH, Inc. (CSI page 1)

  Foreseeability, Premises, Liability, Crime Search, Inc. 281-856-8858. Bring Crime Out of the Dark Into the Light-A Nationwide Crime Research and Analysis Firm. Richard E. Pickard, Expert Witness and Co.

  http://www.crimesearch.net/

- Speech by Attorney General Janet Reno on Crime Mapping
  http://www.usdoj.gov/archive/ag/speeches/1998/agspeech.htm

- The Use of Computerized Mapping in Crime Control and Prevention Programs.
  http://www.ncjrs.org/txtfiles/riamap.txt

- National Institute of Justice Journal
  http://www.ojp.usdoj.gov/nij/journals/jr000241.htm

- Crime Analysis Services Daily Crime Summary
  The Crime Summary-The Crime Analysis Unit publishes a daily.
  http://www.rbpd.org/crime_analysis_services.htm

- Crime Analysis
  Publications and Products: Crime Analysis Ada County, Idaho Michigan State University Crime Analysis for Problem Solvers.
  http://www.usdoj.gov/cops/cp_resources/pubs_prod/s18.htm

- "Computer Mapping a Proven Tool to Fight Arson"
  Fire & Arson Investigator magazine, Sept., 1996. William Lutz, a crime analyst wrote this article about the community of Camden, NJ.

- "New York City Crime Drops 38 Percent" (Article about plotting)
  Government Technology Magazine, March, 1997.

- "Technology puts crime program on the map" (Article by Christina Court discusses plotting.)
  American City & County, August, 1999.

- "Crime Mapping Principle and Practice." (Article by Keith Harris discusses mapping.)
  Us Department of Justice Office of Justice Programs, December, 1999.

- "Geographic Profiling"
  CRC Press, December, 1999.
  Written by Dr. Rossom, a detective with the Vancouver Police Department assigned to the Geographic Profiling Section.

- "A New Pattern For Prevention"

  Peter Combs in the August, 2000 issue of Fire Chief Magazine discuss the Kent Fire Brigade program.

Training

- A.C.T. Now, Inc. (Analysis Consulting and Training Now, Inc.)

  About A.C.T. Now, Inc. Classes offered, Schedule of classes, meet the instructors...

  http://www.actnowinc.org

- Alpha Group Center for crime and intelligence analysis training Instructor's Certificate.

  http://www.alphagroupcenter.com/

- NLECTC Rocky Mountain - Crime Mapping and Analysis Program (CMAP)

  Deputy Director, NLECTC - Rocky Mountain Region Director, Crime Mapping and Analysis Program

  http://www.nlectc.org/cmap/

- Crime Analysis Training Resources (Mass. Association of Crime Analysts)

  Massachusetts Association of Crime Analysts Training Looking to bolster your knowledge of the crime analysis profession?

  http://www.macrimeanalysts.training.com

- Situation report on Crime Analysis at NCIS, Sweden

  The introduction of the Violent Crime Linkage Analysis System.

  http://www.bka.de/text/nb_analytikertagung/situationsbericht_schweden.html

- FVTC - Crime Analysis

  Functions and phases of the crime analysis process. Check for crime analysis training dates.

  http://www.fvtc.edu/

- NIBRS News

  Crime analysis training and crime fighting efforts.

  http://www.asucrp.org/news/index.html

- GPDCA1 Crime Analysis main page

  Analysis, crime definitions, geographic information, report areas, crime information.

  http://www.ci.garland.tx.us/police/GPDCA1.HTM

- Sacramento Police Department Home Page

  Crime reporting, crime prevention, crime statistics, crime alerts Access resources such as alerts and announcements, crime prevention tips, home security notes, missing persons notices, and identity-theft aids.

  http://www.sacpd.org/

- Decatur Illinois Crime Analysis Division

  Decatur Police Department Crime Department and Crime Analysis Unit

  http://www.ci.decatur.il.us/

- Special Units Sex Offender Crime Analysis

  Gilbert Police Department-Crime analysis, detailed crime, statistics, and calls.

  http://www.ci.gilbert.az.us/police/crime.htm

- Scottsdale Police Department's Crime Analysis Unit (CAU)
  What is Crime Analysis? UCR/Part I Crimes
  http://www.ci.scottsdale.az.us/police/Community/CrimeAnalysis/
  cauindex.asp

- CODEFOR - Crime Analysis Unit
  Crime Analysis Unit CODEFOR Police Department Divisions
  http://www.ci.minneapolis.mn.us/citywork/police/about/
  codefor/crime-analysis.html

- Easthampton's Crime Analysis Unit
  http://www.easthamptonpolice.com/cau.htm

- Hampton Police Division Crime Analysis Unit page
  http://www.hampton.va.us/hpd/

- City of Redlands, California, Crime Analysis and Stoppers
  http://www.alphais.com/redlands/3067.html

- Fort Lauderdale Police Department - Crime Analysis Unit
  Finally, the Crime Analysis Unit provides support...
  http://ci.ftlaud.fl.us/police/crimanal.html

- Crime Analysis/Intelligence
  http://www.portsmouth.va.us/ppd/crimeanalysis.htm

- State Troopers Association of Nebraska—Violent Crime Analysis:
  http://www.netroopers.org/texis/scripts/vnews/newspaper/+/ART
  /2000/05/05/3912cfff2

- Miami Beach Police - Crime Analysis Section
  http://www.ci.miami-beach.fl.us/mbpolice/crimeanalysis.htm

- Burnaby RCMP Crime Analysis
  http://www.city.burnaby.bc.ca/rcmp/statistics.html

- Santa Cruz County Sheriff-Crime Analysis
  http://www.scsheriff.com/crimeanalysis.html

- The Daytona Beach Police Department - Crime Analysis
  http://www.ci.daytona-beach.fl.us/police/crime_analysis.htm

- Virginia Beach Police Department Official Website
  http://www.vbgov.com/dept/police/

- Crime Analysis for the Springfield Police Department
  http://www.ci.springfield.mo.us/spd/

- Lake County, IL, Crime Analysis
  http://www.co.lake.il.us/sheriff/crimean.htm

- Houston Police Online
  Crime Analysis, Planning, and Research
  http://www.ci.houston.tx.us/department/police/crime_analysis.htm

- Background on investigative support and crime analysis.
  http://caag.state.ca.us/unsolved/iscau.htm

- LAPD Crime Analysis Section
  http://www.lapdonline.org/organization/icb/crime_section/crime_section.htm

- Kingsport Police Department, Crime Analysis
  http://home.naxs.com/kpdweb/analysis.htm

- Washington County (Oregon) Sheriff's Office - Crime Analysis
  http://www.co.washington.or.us/cgi/sheriff/lec.pl.htm

- Stillwater PD Crime Analysis
  http://www.stillwaterpolicedept.org/CrimeAnalysis.htm

APPENDIX D

Fire Tracking Request Form

PECONIC COUNTY ARSON TASK FORCE
FITS REQUEST FORM

DATE:_____   ASSIGNED CASE #_____

REQUESTED BY:

NAME_____ AGENCY_____

Information required for a Fire Incident Tracking System Report
Probable location of incidents (community, city, township, region, fire, school, or police district) _____
_____
_____

When activity was first noted (approximate date, month, season, and/or year that incidents began) _____
_____

Type of incident noted (rubbish, grass, wildland, dumpsters, mail boxes, vehicles, boat, and/or structures) _____
_____

Approximate number of incidents noted _____

Are there any case numbers assigned to identified incidents?   Yes _____No_____

If yes what are they (attach any field or incident reports that are available to this form)?
_____

Who first noted the activity? _____

Who will be the Point of Contact for information developed?_____
Name _____ Phone # _____
_____
_____

In the space below state any additional information or specific requirements that should be in the report. Please do not state a suspect's name or description unless you wish for the information to be included in the report. If you require additional space use the back of this form._____
_____
_____
_____

Please return this completed form (via hand delivery or secure mail) to the Peconic County Arson Task Force Office.

APPENDIX E

Generic Feedback Request Form

GENERIC FEEDBACK FORM

Page 1 of 2

TO: _____ AGENCY: _____

FROM: _____ DATE: _____

This is a request for feedback on the Fire Incident Tracking Report that your agency recently requested. Please fill out this form, circle your answers where applicable, and return it to the fax number supplied. Your assistance in this matter will help to improve and maintain the quality of future reports. Thank you for your assistance.

Did you receive the report in a timely fashion?   Yes _____   No _____

Was the report what you expected?   Yes _____   No _____

Was the information provided useful in your investigation?   Yes _____   No _____

If not, state why? _____

_____

_____

_____

_____

_____

What was the report used for? Please indicate if there was more than one use for the report.

Establishing patrols _____
Identifying suspects _____
Suspect interview _____
Prevention of future incidents _____
Court case preparation _____

Would you use this process in the future or recommend it to others? Yes _____ No _____

Have there been any additional incidents since the report?   Yes _____   No _____

If yes, please provide a list with date, time, type, and location.

_____

_____

_____

_____

Multiple Fire Setters

GENERIC FEEDBACK FORM

Page 2 of 2

TO: _____ AGENCY: _____

Were there any arrests made in the case?   Yes _____   No _____

If yes, was the track consistent with the suspect's activity?   Yes _____   No _____

If no, what was incorrect?_____
_____
_____
_____
_____

Was this report used by any units and/or agencies other than your own?   Yes _____
No _____

If yes, please list them.
_____
_____
_____
_____
_____

What information did this report lack, if any, and what improvements would you recommend?
_____
_____
_____
_____
_____
_____
_____

Appendix E:
Generic Feedback Request Form

APPENDIX F

Feedback Request Form for a Tracking Report Recipient

PECONIC COUNTY ARSON TASK FORCE
FITS REQUEST FORM

Page 1 of 1

DATE: _____ ASSIGNED CASE # _____
TO: _____ AGENCY: _____
FROM: _____

We would appreciate any feedback on the fire incident tracking report that your agency recently requested. Please fill out this form, circle your answers where appropriate, and return it to the fax number supplied. Your recommendations will help to improve and maintain the quality of future reports. Thank you for your assistance.

Was the information provided useful in your investigation?   Yes _____   No _____

If not, please state why not? _____
_____

Have there been any additional incidents since the report?   Yes _____   No _____

If yes please provide a list with date, time, type, and location._____
_____

Were there any arrests made in the case?   Yes _____   No _____

If yes was the track consistent with the suspect's activity?   Yes _____   No _____

If no what was incorrect?_____
_____

Was this report supplied to any units and/or agencies other then your own? Yes ___ No ___

If yes, please list them._____
_____
_____

Did the maps provided accurately represent the area of concern? Yes _____   No _____

Was there enough detail in the maps provided?   Yes _____   No _____

Were the maps too busy or cluttered with information?   Yes _____   No _____

Did you require additional maps beyond what was provided?   Yes _____   No _____

What information did this report lack, if any, and what improvements would you recommend?_____
_____
_____

APPENDIX G

Short and to the Point Bulletin

MUNICIPAL POLICE BULLETIN
"CONFIDENTIAL"

The following is a notification to be on the lookout for suspicious activity in and around the areas of the 300 to 400 blocks of New Amsterdam Ave. Numerous fires have occurred in a four-block radius of this area. Incidents have occurred from 2000 to 0230 the following day. The most frequent days are Mondays and Wednesdays. The fire activity has included garbage cans, automobiles, and rubbish fires up against buildings. Residential trash pickup occurs on the days noted. Activity suggests the subjects travel by foot and may be known to the area. Note any structures in this area that appear to be abandoned, unoccupied, or have easy access to the public. Any locations noted should be routinely checked. Report any unusual activity or subjects to the 23rd Precinct Crime control Unit.

This message sent by the authority of Lt. E. Springer, 23rd Prct. CCU. Date: 03/20/05. Time: 1600.

Appendix G:

Short and to the Point Bulletin

APPENDIX H

Tip Sheet Used for Multiple Fire Setter Tracking

NEWTON POLICE DEPARTMENT
CRIMINAL INTELLIGENCE UNIT
TIP SHEET

This is for Police Use Only.

Subject: Fire activity
Location: Citywide

Date of notification: 4/25/02
Originated: Det. Nagy - Intel.

Since 12/01 a pattern of intentionally set fires has been noted at abandoned structures throughout the city. This activity has occurred primarily on weekends between the hours of 2300 to 0400 the following day. The unknown suspect has used gasoline at all the fires to accelerate the burn and has targeted entry points off the main road (rear and side entrances). Reports of a late model dark blue or black dodge pick-up have been reported in and around the fire scenes. The activity appears to have originated in Mattington Heights during 8/01, during which time an increase in nuisance fires was noted.

Take note of the following:
• Watch all abandoned and unoccupied structures in your sector.
• Look for subjects carrying containers (of one gallon or more) in and around any probable target structures in your sector.
• Report any vehicles cruising areas in and around probable targets during specified hours and specified days.

Units patrolling the Mattington Heights area should pay special attention to vehicle traffic meeting the description given above during the specified times and days.

Report all activity to Det. Nagy at Intel. or Det. Rose at Arson.

Appendix H:
Tip Sheet used for Multiple Fire Setter Tracking

APPENDIX I

Memorandum to Alert Fire Department Personnel of MFS Activity

METRO FIRE DEPARTMENT

To: All shift Commanders, Batt. Chiefs and Company Officers.
From: J. Martin, Chief of Department.
Date: 4/25/02

This is for internal Fire Department eyes only.

The Fire Investigation division has asked that all command level personnel pay special attention to the Mattington Heights area. Since 12/01 a pattern of intentionally set fires has been noted at abandoned structures throughout the City. This activity has occurred primarily on weekends between the hours of 2300 to 0400 the following day. The unsub has used gasoline at all the fires to accelerate the burn and has targeted entry points off the main road (rear and side entrances). The activity appears to have originated in Mattington Heights during 8/01, during which time an increase in nuisance fires were noted. Any unit responding to fires in known abandoned structures in this area should use extreme caution.

Firefighters can expect fast moving fires with imminent collapse possible. Companies assigned to targeted areas should identify structures matching the criteria and attempts should be made to inspect these probable targets for any homeless or other subjects who frequent the structures in question. All second due officers should attempt to take note of any subjects or civilian vehicles that are frequently sighted at these incidents.

During fire ground operations at these targeted sites, have firefighters pay special attention to any containers that appear to have been recently placed in and around the entry points. Firefighters should also take note of all points of entry, to see if they were secure or forced open prior to their arrival. Report all findings to the lead fire investigator and the chief of the department's office.

APPENDIX J

Generic Crime Mapping Analysis

GENERIC CRIME MAPPING ANALYSIS
FIRE INCIDENT TRACKING SYSTEM REPORT FITS

CONFIDENTIAL BULLETIN   PAGE 1 OF PAGES 3

REPORT ORIGINATED BY M. Martinez FRES	REPORT # 001	COMPLAINT NO. 04-010101	DATE OF THIS REPORT 1/30/02	
TOWNSHIP Islip	SCHOOL DISTRICT Central Region	HAMLET Littleville	FIRE DEPT. Littleville	TYPE OF INCIDENTS auto, debris/ brush, dump, struct
PRECINCT 3rd Police	SECTORS 307&308	DATE OF NOTIFICATION 12/31/03		AGENCIES NOTIFIED Arson Task. Force, County, Police, State

The analysis of more than 54 incidents occurring in the Littleville Fire District, from 2/15/02 through 1/1/04 have revealed the following patterns:

NOTE: The information that follows is not a profile, but a study of the facts surrounding these 7 incidents. These facts include the time, date, day of week, and location of the incidents. The information in this report has been gathered from three independent data sources—the County Fire Rescue Communications Log, the Littleville Fire Dept. incident report database, and the County Police Department Third Precinct Blotter.

FACTS: Fires in the Littleville Fire District are concentrated into three clusters. The clusters in question have been broken up into three separate areas. Refer to the attached map for specific details. The areas have been identified as follows.

1. Railroad depot (Rt. 25) near Norton St. and 10th St.
2. Walter's Elementary School area, near South 21st St.
3. Kentucky Ave. area (Rt. 347) around the end of the tracks (to the north) and railroad station parking (to the south).

Of the 54 incidents the majority occurred in the Littleville Fire District. Fires in the neighboring Fire Districts of Middleton and Burger do not show any patterns or concentration similar to Littleville. The incidents researched involve autos, brush, debris, dumpster, and structure fires. All structure fires in question have occurred after dumpster fires in the same area (approximately 2 to 3 hours later) in the railroad depot area. Highest frequency of incident occurs on weeks of winter and spring break of the Central Region School District. No fire activity was noted on days with precipitation except for activity near Walter's Elementary and in all three cases the activity occurred on Saturdays during the early evening hours of the perceptions days.

As for specific days the analysis revealed the busiest days appears to be Fridays, Saturdays and Sundays with auto fires being the prime target in the Kentucky Ave. area. This would include multiple incidents each day. As for timeframe, fire activity began around 2300 and continued until 0300. A MFA call occurred at 0300 on 11/14/01 at a pay phone in the railroad depot parking area. This was consistent with other incidents that occurred in the early morning hours along the Kentucky Ave. area during 2002. The caller was said to sound drunk.

Multiple Fire Setters

GENERIC CRIME MAPPING ANALYSIS
FIRE INCIDENT TRACKING SYSTEM REPORT FITS

CONFIDENTIAL BULLETIN   PAGE 2 OF PAGES 3

At the **Walter's Elementary School** area, the activity also occurs on weekends during the afternoon and early evening hours (around 1830). The primary targets in this area were brush and debris fires. In the **railroad depot** area structure fires had the smallest time-frame window for occurrence. Starting after 1945 to 0030, primarily on Fridays into Saturdays. Only three structure fires were noted and all occurred in the **railroad depot** area with the majority of targets being dumpster fires. The dumpster fires occurred on Thursday, Friday, and Saturday nights.

CONCLUSIONS:

The targets analysis projects activity to continue to occur on the following days.

Thursdays – Look for incidents to occur between the hours of 1900 to 2359. Activity should focus on areas around the **Railroad depot.** Probable targets are dumpsters.

Fridays – Look for incidents to occur between the hours of 1900 to 2359. Activity should focus on areas around the **railroad depot** and the **Kentucky Ave.** area. Probable targets are dumpsters near the **railroad depot** with a potential for structure fires to occur approximately two to three hours after the original dumpster fire. In the **Kentucky Ave.** area look for automobile fires with a potential for multiple incidents per night.

Saturdays – This is the most active day. Look for incidents to occur between the hours of 0000 to 0400 and again at 1800 to 2359 in all three areas. Activity will begin at 0000 in the **Kentucky Ave.** and the **railroad depot** areas. Activity should shift to the **Walter's Elementary School** area starting around 1600 to1800 with brush and debris fires being the primary targets. Activity will begin to shift back to the **railroad depot** and the **Kentucky Ave.** areas at 2000. Probable targets are dumpsters near the **railroad depot** with a potential for structure fires to occur approximately two to three hours after the original dumpster fire. In the **Kentucky Ave.** area look for automobile fires with a potential for multiple incidents' per night.

Sundays – Look for incidents to occur between the hours of 0000 to 0400 in the **Kentucky Ave.** area with automobile fires as the primary target. From 1300 to 1900, activity should focus on the **Walter's Elementary School Area** starting with brush fires as the primary target.

Appendix J:
Generic Crime Mapping Analysis

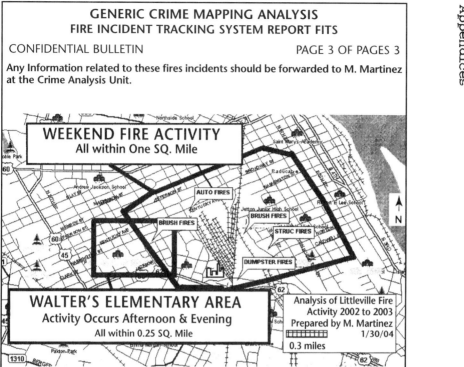

APPENDIX K

Recruitment Form for Volunteer Applicants of a Crime Analysis Unit

PECONIC COUNTY TASK FORCE
FITS UNIT
FIRE INCIDENT TRACKING SYSTEM UNIT
APPLICANT FORM

PAGE 1 OF 9

Thank you for your interest in our unit. As a member of the FITS Unit you will be making a positive contribution to your community.

Entry into the FITS Unit is a very serious matter and that requires extensive application procedures. The application form that you have just received must be completed in a careful and accurate manner. Please be sure to follow the procedures below:

Read and follow all directions.

Leave no blanks (enter N/A for any questions that do not apply to you).

Keep all forms together and do not omit any pages.

If you make an error, cross out the mistake and initial the correction. Do not use white-out or permanent markers.

Once all items are completed, the application must be signed in the presence of a witness, who will also sign the application.

The "Neighbor Form" must be filled out by the three adjacent neighbors to your home.

If you have any questions or are unsure how to fill out the application, please contact the Arson Task Force or leave the item blank until you arrive at Task Force Headquarters.

Be sure to bring the following items with you at the time the application is completed and delivered to Task Force Headquarters.

- High school diploma, GED, and college transcript (if applicable)
- Copy of birth certificate
- Proof of citizenship

PECONIC COUNTY TASK FORCE
FITS UNIT
FIRE INCIDENT TRACKING SYSTEM UNIT
APPLICANT FORM

PAGE 2 OF 9

Applicants may be rejected for intentionally making false statements in this document. Any fraud or deception can lead to criminal charges. Please read all questions carefully and complete all questions.

THIS QUESTIONNAIRE MAY BE PRINTED OR TYPED. ALL ENTRIES, EXCEPT THE SIGNATURE, MUST BE PRINTED LEGIBLY IN BLOCK LETTERS. ENTRIES MUST BE MADE IN BLACK INK. IF ADDITIONAL SPACE IS REQUIRED, USE THE BACK OF EACH PAGE AND INDICATE WHERE ADDED INFORMATION IS FROM BY QUESTION NUMBER.

1. Name _____
     LAST   FIRST   MIDDLE

2. Address   _____

     _____
     TOWN   COUNTY   STATE   ZIP CODE

Home Telephone No. _____

Business Telephone No. _____

3. Give any other names you have used or been known by, and attach a statement giving reason. {if none, so state). _____

4. Where were you born? _____

   CITY OR TOWN OF BIRTH_____

   COUNTY, STATE, and COUNTRY_____

5. BIRTH CERTIFICATE NUMBER_____

6. DATE OF BIRTH_____

   SOCIAL SECURITY NUMBER _____

7. Are you native born or a naturalized citizen?{state which) _____

8. If naturalized, indicate details on back of Page 1.

9. List all places where you registered or voted. {If none, state so on line below.)
   RESIDENCE AT TIME (City or town, County, State) _____

   _____

   _____

Candidate's Signature_____

Print Your Name Here _____

224

PECONIC COUNTY TASK FORCE
FITS UNIT
FIRE INCIDENT TRACKING SYSTEM UNIT
APPLICANT FORM

10. In CHRONOLOGICAL ORDER, starting with the present and working to the past, list each and every place in which you have resided since you left elementary school:

FROM MONTH YEAR	TO MONTH YEAR	ADDRESS OF RESIDENCE	BOROUGH OR COUNTY	CITY OR TOWN	STATE	ZIP

11. List all the schools and colleges you have attended:

FROM MONTH YEAR	TO MONTH YEAR	SCHOOL	DAY OR EVENING	EXACT ADDRESS	LAST GRADE OR TERM

12. What college degree or professional licenses do you possess? List High School Equivalency Diploma Number. If applicable and date:

13. Give the name of your father, mother (maiden name), wife, husband, sisters, brothers, aunts, uncles, fathers or mothers in-law:

RELATIONSHIP	NAME	DOB	ADDRESS	OCCUPATION	LIVING/DEAD

14. ARREST- has any of the above listed individuals ever been arrested? I yes, give details on blank pages, include names, dates of birth, and relationship.

15. What is your current occupation or calling?_____

16. Are you now the owner, partner, stock holder, or corporate member in any business? ☐Yes   ☐No If answer is yes, give details on seperate sheet.

Candidates Signature_____

Print Name_____

Recruitment Form for Volunteer Applicants of a Crime Analysis Unit

PECONIC COUNTY TASK FORCE
FITS UNIT
FIRE INCIDENT TRACKING SYSTEM UNIT
APPLICANT FORM

PAGE 4 OF 9

CHRONOLOGICALLY, starting with the present and working to the past, list employment, unemployment, military time (keep in sequence) including all part-time employment, and any "off the books" employment.

FROM Month/Year	TO Month/Year	Name of Employer (Work Site, Zip,Tel.)	Employer Address	Position Held	Reason for Leaving	Supervisor

17. If spouse is working, list:

Occupation:_____ Employer::_____

Address_____ Telephone:_____

18. Were you ever discharged or asked to resign from employment? ☐Yes ☐No How many times?_____
Give the details of discharges or forced resignation below:

EMPLOYER	EMPLOYERS ADDRESS	DATE	SUPERIOR	REASON/ DISCHARGE

Candidates Signature_____

Print Name_____

Multiple Fire Setters

PECONIC COUNTY TASK FORCE
FITS UNIT
FIRE INCIDENT TRACKING SYSTEM UNIT
APPLICANT FORM

PAGE 5 OF 9

19. MILITARY INFORMATION

Branch of Service_____ Date Entered:_____

Rank held upon discharge_____ Serial Number:_____

Active Service: from:_____ to:_____

TYPE OF DISCHARGE: ☐Honorable ☐Other

An other than Honorable Discharge is not an absolute bar to employment.

Have you had any disciplinary action while in the Service? ☐Yes ☐No (if yes, list on seperate sheet)

RESERVE/OBLIGATION: Active/Inactive From:_____

To:_____

Address entered

from:_____

Place of

Discharge:_____

Address returned to:_____

20. Are you registered for the Draft? ☐Yes ☐No

Have you ever been rejected by the Military prior or after induction? ☐Yes ☐No

If yes, reason:_____

_____

Selective Service Number:_____ or (Local Board) place of

Registration:_____

Address:_____

Last classification:_____ Date Classified/Registered:_____

21. Were you ever arrested or taken in to custody, in this state or elsewhere (include all investigations, even those as a juvenile)? ☐Yes ☐No If yes, how many times?_____

Indicate below all arrest, including juvenile, youthful offender, and wayward minor (sealed records included):

DATE	VIOLATION Actual Charge	LOCATION	CHARGE Reduced to	Court Disposition or Sentence	Police Agency Concerned

Candidates Signature_____

Print Name_____

Appendix K:

Recruitment Form for Volunteer Applicants of a Crime Analysis Unit

PECONIC COUNTY TASK FORCE
FITS UNIT
FIRE INCIDENT TRACKING SYSTEM UNIT
APPLICANT FORM

27. Did you ever receive complaints from a Police Department or any other regulatory authority in connection with any employment? ☐Yes ☐No If yes, give details:_____

28. Have you, or any corporation or partnership of which you were an officer, director or partner, ever possessed a license or permit, (exclude driver's license or lerner's permits) issued by any governmental agency? (include pistol license, liquor license, etc.) If yes, give details, if no, state so:_____
_____
_____

29. Have you ever been authorized to collect a State Sales tax? If yes, give State and authorization #:_____

30. Has any license or permit, (exclude driver's license or learner's permits) issued by any City, State, or Federal Agency ever been denied to you or to any corporation or partnership of which you were an officer, director, or partner? has any such license or permit ever been cancelled or suspended? If yes give details, if no state so:_____
_____

31. Have you ever acted as a sponsor, character witness or made any recommendations for, or concerning any person or premises, to any municipal, state, or federal agency in connection with the issuance, revocation, or suspension of any license, or permit, or for any other reason, for any other person or premises? ☐Yes ☐No
If yes, give details:
_____

32. Have you ever submitted application for a Civil Service position? ☐Yes ☐No If yes, complete the chart:

YEAR	POSITION	# ON LIST	FEDERAL, STATE MUNICIPALITY	INVESTIGATED, YES, NO IF YES, GIVE INV. NAME	ACCEPTED/ REJECTED	REJECT REASON

33. Have you ever previously submitted an application to any Police Department? ☐Yes ☐No
If yes, state year:_____ Department:_____

34. Have you ever been fingerprinted? ☐Yes ☐No
When?_____ Where?_____
Purpose?_____

When?_____ Where?_____
Purpose?_____

35. List name and address of current spouse and all formal spouses:
_____
_____

36. Are you living with your spouse? ☐Yes ☐No

37. If ever separated, annulled, or divorced, indicate which below, and fill in required information:

Seperately Annulled or Divorced (Indicate Which)	Date Issued	By Whom	Where Issued (Court and State)	Offending Party as Decreed by Law	Reason

Candidates Signature_____

Print Name_____

PECONIC COUNTY TASK FORCE
FITS UNIT
FIRE INCIDENT TRACKING SYSTEM UNIT
APPLICANT FORM

38. List below all your children:

NAME	D.O.B.	PLACE OF BIRTH	WITH WHOM AND WHERE DOES CHILD RESIDE

39. Are you now supporting all children born to you, adopted, and stepchildren? ☐Yes ☐No If yes, give full details:_____

40. Have you ever been involved as a defendant in a paternity proceeding? ☐Yes ☐No If yes, give full details:_____

41. Have you ever been bonded? ☐Yes ☐No With respect to each time bonded, state details below:

REASON	BY WHOM, NAME AND ADDRESS	DATE

42. Have you ever been refused a Bond? ☐Yes ☐No If yes, by whom?_____

43. Have you ever by word of mouth or in writing, advocated the doctrine that the Government of the United States of America or of any State or of any political subdivision thereof should be overthrown or overturned by force, violence, or any unlawful means?_____

44. Are you currently using or experimenting with, to any extent, any drugs, narcotics, or controlled substances, including marijuana and it's derivatives? _____

45. Have you ever sold, given away, or profited from selling any substance listed as an unlawful controlled substance in any state or federal statute? _____

46. Have you ever engaged in any illegal gambling activities?_____

47. Have you ever taken a polygraph (lie detector) examination?_____
Where?_____ Date?_____

48. Do you have any objections to taking a polygraph prior to appointment as an Auxiliary Officer?_____

Candidates Signature_____

Print Name_____

PECONIC COUNTY TASK FORCE
FITS UNIT
FIRE INCIDENT TRACKING SYSTEM UNIT
APPLICANT FORM

PAGE 8 OF 9

49. Do you have a physical or mental disability that would prevent you from performing in a reasonable manner the duties required of this position?_____

Do you have any knowledge or information, in addition to that called for in the preceding questions, that is or may be relevant, directly or indirectly, in connection with an investigation of your eligibility or fitness for the position of_____including, but not limited to, knowledge of information concerning your character, physical or mental condition, temperance, habits, employment, education, subversive activities, family, associations, criminal records, traffic violations, residence, or otherwise?　☐Yes　☐No　If yes, give details.

_____

_____

_____

_____　_____

CANDIDATE SIGN HERE

SWORN TO ME, THIS_____ DAY OF _____ 20_____

_____

(NOTARY PUBLIC OR COMMISSIONER OF DEEDS)

Appendix K:

Recruitment Form for Volunteer Applicants of a Crime Analysis Unit

Multiple Fire Setters

PECONIC COUNTY TASK FORCE
FITS UNIT
FIRE INCIDENT TRACKING SYSTEM UNIT
APPLICANT FORM

Each_____ applicant must supply the names and addresses of 3 immediate neighbors. Daytime phone numbers for each should e included if known. This sheet must be completed and returned with the completed_____ candidate's investigative questionnaire.

1.Name:_____ Address:_____

2.Name:_____ Address:_____

3.Name:_____ Address:_____

Appendix K:

Recruitment Form for Volunteer Applicants of a Crime Analysis Unit

Glossary

AFA *Automatic Fire Alarms* is an activation of a fire detection system. The types of activations will vary but the most common are faulty smoke detectors, electrical wires grounding (causing a short in the circuit), drop in water pressure followed by a rapid rise causing a water flow activation, and the accidental trip of a manual pull station. As with the different types of alarm activation's, the fire alarm systems will vary in devices and technology. If you have no background with these systems do not attempt to learn about them. Instead consult an expert on them who can answer your questions. Good sources are your local Fire Prevention or Inspection Officer or industry experts such as Underwriters Laboratory (UL).

AIMS *Arson Information Management System* is computer software developed by Fire Investigators at the National Fire Academy. It is designed as a clearinghouse for key information related to fire incidents. The information is entered locally by investigators at their convenience and stored in a database. The database can be searched by entering key phrases, words, names, times and/or dates; the database will retrieve any items similar to the item entered.

APRS   *Arson Pattern Recognition System* developed for the Tennessee State Fire Marshal's in 1975 and operated by the University of Tennessee.

ANN   *Artificial Neural Networks* is a term used to describe computerized Artificial Intelligence systems. These systems have the ability to independently resolve issues and solve problems using a network of interconnected processors.

ATF   *Federal Bureau of Alcohol, Tobacco, and Firearms* is the federal law enforcement agency with jurisdiction over fires and bombings involving interstate commerce.

FBI   *Federal Bureau of Investigation* is the federal law enforcement agency with jurisdiction over fire and bomb related terrorist and hate (bias) crimes.

CAD   *Computer Aided Dispatching*—many municipalities now have these system. They will vary with each jurisdiction, but this could be one of the best sources for tracking information.

CMRC   *Crime Mapping Research Center* is the Federal agency tasked to study and disburse knowledge of grant information and software developments in the area of crime mapping. This agency falls under the direction of the National Institute of Justice which are over seen by the Department of Justice.

CI's   *Confidential Informants* is the term used for a human intelligence source whose identity cannot be revealed in order to maintain their anonymity.

DOJ   *Department of Justice* headed by the Attorney General is the federal agency that covers the federal law enforcement agencies of the FBI, DEA, as well as many Law Enforcement research groups including the National Criminal and Justice Research Services, National Institute of Justice, Office of Juvenile Justice and Delinquency Prevention, Office of Victim of Crime, Bureau of Justice Statistics, Bureau of Justice Assistance, and the Office of National Drug Control Policy.

FCC *Federal Communications Commission* is the agency who has authority over the use of radio frequencies and provides the licensing to broadcast on those frequencies. Their offices are located in Washington, DC, Emmitsburg, MD, and regionally through out the USA.

FEMA *Federal Emergency Management Agency* is the parent agency of the National Fire Academy (NFA). FEMA provides training for public safety organizations and grants for programs related to disasters and mitigation. FEMA can be a good source for grant proposals.

FITS *Fire Incident Tracking* is the name used by the author to identify the final product of the Tracking Process. FITS may also be thought of as an intelligence report or Tip sheet.

GIC *Good Intent Calls* are incidents that are reported, as emergencies, but do not require the fire suppression response. For example a report of smoke coming from a house across from the caller's residence turns out to be smoke from the homes chimney.

GIS *Geographic Information Systems* are computer based software systems designed to allow for the analysis of data by location and revealing hidden patterns, relationships, and trends that are not readily apparent in conventional data. This process is based on the geocoding (assigning an X and Y coordinate to an address) so that data can be placed on a map. This is done by assigning street centerlines or latitude and longitudes (per property parcels) to location. The data is then related to the locations through a database. Many governments will have GIS specialists. Check with them, this could be the source for a wealth of information to help you get started.

GPS *Global Positioning System*—this is a constellation of satellites designed to give the operator the ability know where they are at anytime anywhere in the world. It will give you the latitude, longitude, and altitude instantly with in a few feet of your actual position. This can be very useful when documenting and tracking wildland fire incidents.

Geo. Profile *Geographic Profiling* is an investigative technique for apprehending suspects that focuses on the geographic environment and considers how different criminals and criminal acts utilize geographic space differently. With this technique and understanding, law enforcement officers can better design a search strategy and identify high probability areas of further criminal activity.

HITS *Homicide Investigation and Tracking System* is a computer based data system used to conduct a linkage analysis of serial crime cases. Developed for the Washington State Attorney General's Office, the system has been in service since 1987.

HI *Human Intelligence* refers to individuals sources that provide information such as tips or a lead on incidents and/or cases.

IVO *In the Vicinity* is a phrase used to explain a general (or non specific) incident location. This phrase may be found on fire reports or incident data sheets.

meta data This would be the primary information sources from which all other databases are developed. The *meta data* identifies those primary sources.

MFA *Malicious False Alarms* or crank calls is a phrase used to explain an intentional activated alarm or report of false incident. This phrase may be found on fire reports or incident data sheets.

MFRI *Maryland Fire and Rescue Institute* is the firematic training branch of the University of Maryland. They publish a quarterly bulletin and are a good source for training and research.

MFS *Multiple Fire Setters* is the term used in this book to identify any individual who intentionally sets numerous fires over a prolonged period of time (usually more then twenty-four hours).

MIS *Management Information Systems* are the data collection experts' resources in your local jurisdiction. They may also be entitled Information Management System (IMS), data processing, and Global Information Systems (GIS). These agencies could be the source for provide a wealth of information to help you get started.

NCIC *National Crime Information Center* operated by the FBI.

NCJRS *The National Criminal Justice Reference Service* is an online and telephone fax service of information related to criminal and juvenile justice. It is a program run under the U.S. Department of Justice.

NFIRS *National Fire Incident Reporting System*, started by the United State Fire Administration in the 1970s, is a voluntary reporting network.

NFPA *National Fire Protection Association* is an organization made up of private and public agencies concerned with fire and associated fire issues. This association studies fire problems and develops recommended standards to help prevent fires and protect from further issues.

NCAVC *National Center for the Analysis of Violent Crime* formally the FBI Behavioral Science Unit located in Quantico, Virginia. This center is a joint operation of the FBI and ATF in the study of violent crimes. The center develops the profiles of multiple fire setters.

TRO *To the rear of* is the term used in incident reports to identify when an activity occurs behind or at the back of a structure or property.

unsub *Unknown Subject* is term used by Federal and Local Law Enforcement to explain the person or persons currently unidentified in the investigation. In many cases this would be the prime suspect in a criminal investigation.

UCR *Uniform Crime Reporting* is the term given to the Department of Justice national incident report system, which is used by most law enforcement agencies. The data is used for the basis of federal, state, and local law enforcement programs as well as grant funding.

USFA *United States Fire Administration* is an office of the Federal Emergency Management Agency. It provides training and education through the National Fire Academy and funding in the form of grants to the nation's fire services. Their offices are located in Washington, DC, and Emmitsburg, MD.

VICAP *Violent Criminal Apprehension Program* functions as a database for serious violent offenders and unsolved extremely violent crimes. The database is operated by the NCAVC Center of the FBI and ATF.

Bibliography

Agung, A. "Crime Hot Spot Analysis and Dynamic Pin Map." Proceedings, Environmental Systems Research Institute International User Conference, 1997. Available at http://www.esri.com/library/userconf/archive.html.

Baker, K.C., and R. Buettner. "They Set Fires To Buy Snacks." New York Daily News, www.nydailynews.com, 1996.

Barnett, W., and M. Spitzer. "Pathological Fire-setting 1951-1991: A Review." Medical Science Law. Vol.34, No.1, 1994.

Basilesco, J. "Derry Police Hunt for Firebugs." Eagle-Tribune, www.eagletribune.com, 1998.

Beasimer, J. "Police Probing Arsons Detective: 11 Blazes May Be Connected." The Poughkeepsie Journal, www.poughkeepsiejournal.com, 1998.

Bennett, W.D., A. Merlo, and K. K Leiker. "Geographical Patterns of Incendiary and Accidental Fires." Journal of Quantitative Criminology, Vol. 3, No. 1.

Block, C.R, M, Dabdoub, and S. Fregly. "Crime Analysis Through Computer Mapping." Police Executive Research Forum. Washington, DC, 1995.

Boyle, M. "Police Arrest Man in String of Arsons." The Standard-Times, www.standardtimes.com, 1996.

Brantingham, P.J. and P.L. Brantingham. "Patterns in Crime." MacMillian, NY, 1984.

Brown, C.E. "Suspicious Fires Hit Central Seattle Area." The Seattle Times Company, www.seattletimes.com, 1997.

Bruce, C. "Ten Commandments of Crime Analysis." International Association of Crime Analysts, Fall, 1999.

Brush, P. "Serial Arsonist Strikes Boats." APB News, 1999.

Burton, G. "Two Questioned About Arsons." The Salt Lake Tribune, www.sltrib.com, 1998.

Carter, R. E. "Arson Investigation." Glencoe Publishing Co., Encino, CA, 1978.

Chicca, E.L. "Developing Automated Systems to Track False Alarms for Local Jurisdictions." Proceedings, Environmental Systems Research Institute International User Conference, http://www.esri.com/library, 1997.

Coombs, P., and A. Eckley. "New Pattern for Prevention." Fire Chief, 78-80,(August, 2000), 78-80.

DeRoy, B. "Restroom Reeks of Another Arson Fire." The Minnesota Daily, www.mndaily.com/newsite/, 1995.

Douglas, J., and M. Olshaker. "Mind Hunter Inside the FBI's Elite Serial Crime Unit." Simon and Schuster, New York, NY, 1995.

Dwyer, J. "The Devil Watched As Bushwick Burned." New York Daily News, www.nydailynews.com, 2000.

Egbert, B., and M. McPhee. "Man Eyed as Firebug-He may be arsonist who set 18 blazes." New York Daily News, www.nydailynews.com, 2000.

Foster, G. "Playing with Fire: Troubling Crimes in Lilac City." Foster's Online, www.fosters.com, 1997.

Garnham, P. "FEMA Concedes Fire Software Market." Fire-Rescue Magazine, (15-16, December, 1999)., 15-16.

Gaynor, J., and C. Hatcher. "The Psychology of Child Firesetting." Brunner-Mazel, New York, NY, 1987.

Gearty, R. "School Arson Eyed-Science test fear may have fueled student torcher." New York Daily News, www.nydailynews.com, 1999.

Gelston, D. "Stripper sentenced in Church Arsons." the Associated Press, www.abcnews.com, 2000.

Grescoe, T. "Murder, He Mapped." Canadian Geographic, (49-52, September/October), 1996. 49-52.

Hamilton, P.L. "Arson Suspect in Custody." Lubbock Avalanche-Journal, www.lubbockonline.com, 1997.

Hamling, J. "A Psychodynamic Classification System for Pathological Firesetters With Treatment Strategies for Each Subgroup." J.E. Hamling. 1995.

Harries, K. "Mapping Crime: Principle and Practice." National Institute of Justice, U.S. Dept. of Justice, Washington, DC, December, 1999.

Huppe, R.W. "Man Arrested in Church Fires." The Associated Press, www.abcnews.com, 2000.

Hughes, W. P. Jr. "Fleet Tactics Theory and Practice." U.S. Naval Institute Press, Annapolis, Maryland, 1986.

Icove D.J., and P.R. Horbert. "Introduction to the Serial Arsonist: Research by the NCAVC." National Center For the Analysis of Violent Crime, FBI Academy, Quantico, VA. May, 1990.

Icove D. J., and P.J. Keith. "Principles of Incendiary Crime Analysis: The Arson Pattern Recognition System (ARPS) Approach to Arson Information Management." Kentucky Dept. of Public Safety, Knoxville, TN, 1983.

Kalfrin, V. "Texas Town Enacts Curfew to Fight Arson Spree." APB News, 2000.

Kassab, B. "Man Questioned in Arson." www.orlandosentinel.com, Fl, 2001.

Kettler, B. "State Agencies Battle Arsonists." Mail Tribune, www.mailtribune.com, 1997.

Levin, B. "Psychological Characteristics of Firesetters." Fire Journal, Vol. 70, 36-41.

Lewis, N.D.C., and H.Yarnell. "Pathological Firesetting (Pyromania)." Nervous and Mental Disease Monograph, No.82, Coolidge Foundation, New York, NY 1951.

Luo M., and J.J. Palmer. 2000. "Group Claims Role in Fire." Newsday, Melville, NY. 2000.

Lutz, W.E. 1996. "Computer Mapping: A Proven Tool to Fight Arson." Fire Arson Investigator, Vol. 47, No. 1, 15-18.

Mackay, R. "Geographic Profiling: A New Tool For Law Enforcement." The Police Chief, 51-59, (December, 1999). 51-59.

Macko, S. November 17, 1995. "Firefighters Turned Arsonists." Emergency News Network, www.emergency.com, 1995.

Macleod, I. February 21, 1999. "Fire Bug Spreads Fear in West End." The Ottawa Citizen, www.canada.com/ottawa/, 1999.

Mamalian, C.A., N.G. LaVigne, and Staff of Crime Mapping Research Center. 1999. "The Use of Computerized Crime Mapping By Law Enforcement: Survey Results." National Institute of Justice, US DOJ, Washington, D.C,. 1999.

Marlowe, M. January 31, 2000. "Ohio State U Campus Arson Fires Continue." U-Wire, 2000.

Martin, D., E. Barnes, and D. Britt. 1998. "The Multiple Impacts of Mapping It Out: Police, Geographic Information Systems (GIS) and Community Mobilization During Devil's Night in Detroit, Michigan." Crime Mapping Case Studies: Successes in the Field. Police Executive Research Forum. Washington, D.C., Police Executive Research Forum, 3-14. 3-14.

McGarigle, B. 1997. "Crime Profilers Gain New Weapons." Government Technology, December, 1997.

McGraw, S. November 20, 2000. "Convicted Arsonist Eyed in Fires." APB News, 2000.

McNamara, J. December 14, 1997. "The Justice Story: Dangerous Firebug." New York Daily News, www.nydailynews.com, 1997.

Navrot, M. April 20, 1997. "Big Jump in Arson Cases Keeps Heat on Local Firefighters." Lubbock Avalanche-Journal, www.lubbockonline.com, 1997.

Navrot, M. November 18, 1997. "Adams sentenced, Fined for Area Fires." Lubbock Avalanche-Journal, www.lubbockonline.com, 1997.

Noble, I. January 10, 2000. "Arson Link Probed." North Shore News, www.nsnews.com, 2000.

Panel Discussion. 1983. "Using Computers in Crime Fighting, Training and Administration: A Practical Approach." Police Chief, Vol. 50, March, 1983.

Phillips, R.A. April 12, 1999. "A 'Megan's Law' For Arsonists." APB News, 1999.

Pignone, S.A. October 22, 1999. "Woman, Her Children Charged in Arsons."APB News, 1999.

Pratt, M. 2000. "Modeling Fire Hazard" ArcUser, Environmental Systems Research Institute International. (July-September, 2000), 32-39.

Price, J. June 25, 1997. "Night of Several Fires Revives Arson Worries." The News & Observer, www.news-observer.com, 1997.

Raftery, T., and B. Farrell. July 7, 2000. "Accused of Setting Six Gravesend Fires." New York Daily News, www.nydailynews.com, 2000.

Rich, T.F. 1995. "The Use of Computerized Mapping in Crime Control and Prevention Programs." Washington, DC. US Department of Justice, National Institute of Justice, Washington, DC.

Rider, A.O. 1980. "The Firesetter: A Psychological Profile." FBI Law Enforcement Bulletin, June, July, and August, 1980.

Rippel A. C. January 13, 2001. "Judge Trims Bail in Arson Suspect's Case." www.orlandosentinel.com, 2001.

Rossmo, D.K., J.L. Jackson, and D.A. Bekerian. 1997. "Offender Profiling: Theory, Research and Practice, Chapter 9-Geographic Profiling." John Wiley & Sons, NY, (1997), 159-175.

Sapp, A.D., T. Huff, G. Timothy, G. Gary, D.J. Icove. et al. "A Motive-Based Offender Analysis of Serial Arsonist." Fedreal Bureau of Investigation and the Federal Emergency Management Agency. 5-12 and 56-88.

Stauffer, C., and T. Murse. January 25, 2001. "Firefighter Charged in PA Arson." Lancaster New Era, Lancaster, PA, 2001.

Tuma, D. April 24, 2000. "Struggle to Regroup 29 Homeless in Montauk's Latest Fire." New York Daily News, www.nydailynews.com, 2000.

Turvey, B.E. 1998. "Deductive Criminal Profiling: Comparing Applied Methodologies Between Inductive and Deductive Criminal Profiling Techniques." Knowledge Solutions, www.corpus-delicti.com. January, 1998.

Van Pelt, R. December, 2000. "Arrest Closes 6 Suspicious Fires." Fireline, Dutchess County, NY, December, 2000.

Venezia, T. June 11, 1999. "Fire Investigator Charged with Arson." APB News, 1999.

Waterstrat, S. April 10, 2000. "Arson Hits Fraternity Houses—One Blaze Was Set to Kill, Officials Say."APB News, 2000.

Williams, R. 1986. "Computerized Crime Map Spots Crime Patterns." Law and Order, Vol. 110.

Wise, Barry. 1995. "Catching Crooks with Computers." American City and County, Vol. 110, May, 1995.

Worden, A. June 18, 1999. "100 Agents Probe Synagogue Fires." APB News, 1999.

Unknown Author, 1999. "News in Brief: NASA Global Satellite System has Application for Monitoring Wildfires." Fire Engineering, (February, 1999), 24.

Unknown Author, July 15, 1999. "Six Firefighters Charged in Deadly Arson." APB News, 1999.

Unknown Author, July 28, 1999. "Accused Church Arsonist Charged in More Fires." APB News, 1999.

Unknown Author, August 11, 1999. "Arsonists Torch Baton Rouge 24 Times." APB News, 1999.

Unknown Author, October 27, 1999. "Two Charged in 100-Fire Arson Spree." APB News, 1999.

Unknown Author, January 18, 2000. "70 Vacant Homes Torched in Fla. Arson Spree." APB News, 2000

Unknown Author, February 8, 2000. "Senior Resident charged With Deadly Arson." APB News, 2000

Unknown Author, February 26, 2000. "Satan Fan Admits Burning Churches, Say Cops." APB News, 2000.

Unknown Author, July 2000. "Church Arsonist Pleads Guilty Lucifer's Missionary Linked to 26 church Fires." The Associated Press, www.abcnews.com, 2000.

Unknown Author, July 12, 2000. "Missionary of Lucifer Admits 26 Church Arsons." APB News, 2000.

Unknown Author, December 2000. "Firefighters Bust Dumpster Burner." The Long Island Advance, 2000.

Index

D

E

G

H

I

J

K

L

M

N

O

T

U

V

W

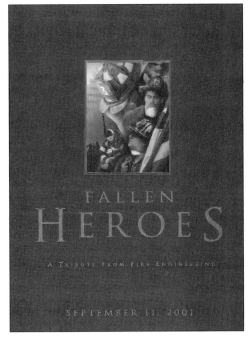

HERE'S WHAT CUSTOMERS ARE SAYING ABOUT SHOPPING ONLINE AT WWW.PENNWELL-STORE.COM:

"The service was great; I had my order within a few days — when all other stores didn't have it in stock."
— Scott R., Accokeek, MD

"I was very pleased with the service. Excellent response to my e-mail inquiring about my order status. I will be ordering from PennWell again in the near future."
— Chester G., Wilmington, DE

"I couldn't find a couple of items, I left an email, and they shipped the items as well. The online store is excellent and has my highest regards and approval."
— Scott E., Ilion, NY

"Being that I haven't ordered online at all in the past, the only basis I had for the quality and speed of service was the feedback from friends and relatives. PennWell has certainly made my first online experience a pleasant one…"
— Hercules R., Westminster, CA

"Already received the order and the invoice — it was quite user-friendly. Will definitely order again online. Thank you!"
— Brenda P., Denver, CO

What are you waiting for? Shop online today at
www.pennwell-store.com!

Don't forget to sign up for our e-newsletter to keep up with our latest titles and offers!